Advances in Industrial Control

Other titles published in this series:

*Digital Controller Implementation
and Fragility*
Robert S.H. Istepanian and James F.
Whidborne (Eds.)

*Optimisation of Industrial Processes
at Supervisory Level*
Doris Sáez, Aldo Cipriano and Andrzej W.
Ordys

Robust Control of Diesel Ship Propulsion
Nikolaos Xiros

Hydraulic Servo-systems
Mohieddine Jelali and Andreas Kroll

*Model-based Fault Diagnosis in Dynamic
Systems Using Identification Techniques*
Silvio Simani, Cesare Fantuzzi and Ron J.
Patton

Strategies for Feedback Linearisation
Freddy Garces, Victor M. Becerra,
Chandrasekhar Kambhampati and
Kevin Warwick

Robust Autonomous Guidance
Alberto Isidori, Lorenzo Marconi and
Andrea Serrani

Dynamic Modelling of Gas Turbines
Gennady G. Kulikov and Haydn A.
Thompson (Eds.)

Control of Fuel Cell Power Systems
Jay T. Pukrushpan, Anna G. Stefanopoulou
and Huei Peng

*Fuzzy Logic, Identification and Predictive
Control*
Jairo Espinosa, Joos Vandewalle and
Vincent Wertz

*Optimal Real-time Control of Sewer
Networks*
Magdalene Marinaki and Markos
Papageorgiou

Process Modelling for Control
Benoît Codrons

*Computational Intelligence in Time Series
Forecasting*
Ajoy K. Palit and Dobrivoje Popovic

*Modelling and Control of Mini-Flying
Machines*
Pedro Castillo, Rogelio Lozano and
Alejandro Dzul

Ship Motion Control
Tristan Perez

Hard Disk Drive Servo Systems (2nd Ed.)
Ben M. Chen, Tong H. Lee, Kemao Peng
and Venkatakrishnan Venkataramanan

*Measurement, Control, and
Communication Using IEEE 1588*
John C. Eidson

*Piezoelectric Transducers for Vibration
Control and Damping*
S.O. Reza Moheimani and Andrew J.
Fleming

Manufacturing Systems Control Design
Stjepan Bogdan, Frank L. Lewis, Zdenko
Kovačić and José Mireles Jr.

Windup in Control
Peter Hippe

*Nonlinear H_2/H_∞ Constrained Feedback
Control*
Murad Abu-Khalaf, Jie Huang and
Frank L. Lewis

Practical Grey-box Process Identification
Torsten Bohlin

Control of Traffic Systems in Buildings
Sandor Markon, Hajime Kita, Hiroshi Kise
and Thomas Bartz-Beielstein

Wind Turbine Control Systems
Fernando D. Bianchi, Hernán De Battista
and Ricardo J. Mantz

*Advanced Fuzzy Logic Technologies in
Industrial Applications*
Ying Bai, Hanqi Zhuang and Dali Wang
(Eds.)

Practical PID Control
Antonio Visioli

(continued after Index)

Jian-Xin Xu • Sanjib K. Panda • Tong H. Lee

Real-time Iterative Learning Control

Design and Applications

 Springer

Jian-Xin Xu, PhD
Sanjib K. Panda, PhD
Tong H. Lee, PhD

Department of Electrical and Computer Engineering
National University of Singapore
4 Engineering Drive 3
117576 Singapore
Singapore

ISBN 978-1-84996-824-9   e-ISBN 978-1-84882-175-0

DOI 10.1007/978-1-84882-175-0

Advances in Industrial Control series ISSN 1430-9491

A catalogue record for this book is available from the British Library

Cover design: eStudio Calamar S.L., Girona, Spain

Printed on acid-free paper

9 8 7 6 5 4 3 2 1

springer.com

Advances in Industrial Control

Professor (Emeritus) O.P. Malik
Department of Electrical and Computer Engineering
University of Calgary
2500, University Drive, NW
Calgary, Alberta
T2N 1N4
Canada

Professor K.-F. Man
Electronic Engineering Department
City University of Hong Kong
Tat Chee Avenue
Kowloon
Hong Kong

Professor G. Olsson
Department of Industrial Electrical Engineering and Automation
Lund Institute of Technology
Box 118
S-221 00 Lund
Sweden

Professor A. Ray
Department of Mechanical Engineering
Pennsylvania State University
0329 Reber Building
University Park
PA 16802
USA

Professor D.E. Seborg
Chemical Engineering
3335 Engineering II
University of California Santa Barbara
Santa Barbara
CA 93106
USA

Doctor K.K. Tan
Department of Electrical and Computer Engineering
National University of Singapore
4 Engineering Drive 3
Singapore 117576

Professor I. Yamamoto
Department of Mechanical Systems and Environmental Engineering
The University of Kitakyushu
Faculty of Environmental Engineering
1-1, Hibikino,Wakamatsu-ku, Kitakyushu, Fukuoka, 808-0135
Japan

To our parents and

*My wife Iris Chen and daughter
Elizabeth Xu*

                       – Jian-Xin Xu

*My wife Bijayalaxmi and daughters
Pallavi and Puravi*

                – Sanjib Kumar Panda

My wife Ruth Lee

                     – Tong Heng Lee

Series Editors' Foreword

The series *Advances in Industrial Control* aims to report and encourage technology transfer in control engineering. The rapid development of control technology has an impact on all areas of the control discipline. New theory, new controllers, actuators, sensors, new industrial processes, computer methods, new applications, new philosophies…, new challenges. Much of this development work resides in industrial reports, feasibility study papers and the reports of advanced collaborative projects. The series offers an opportunity for researchers to present an extended exposition of such new work in all aspects of industrial control for wider and rapid dissemination.

Advances in computer and software technology have allowed a number of control techniques to become feasible in applications. The ability to store and recall performance error trajectory, computed previously, in a process has permitted iterative learning control to be employed in the real-time industrial applications. The type of application that suits this method, is one in which the process is repeated under the same or similar conditions, so that learning from past errors is possible, and the process is relatively free of unpredictable disturbances. The classic exemplar is the repetitive robotic operation present in a manufacturing production line. Here repetition is a natural part of the process operation and can be exploited to achieve improved control.

An early reported application can be found in a *Journal of Robotic Systems* paper by S. Arimoto, S. Kawamura and F. Miyazaki (1984) titled, 'Bettering operation of robots by learning', however the *Advances in Industrial Control* series can claim the distinction of publishing the first monograph on this new control technique with K.L. Moore's *Iterative Learning Control for Deterministic Systems* (ISBN 978-3-540-19707-2, 1993). Various other books have been published in the meantime, including a more theoretical volume by J.-X. Xu and Y. Tan, entitled *Linear and Nonlinear Iterative Learning Control* (ISBN 978-3-540-40173-5, 2003). This *Advances in Industrial Control* volume, *Real-time Iterative Learning Control* by Jian-Xin Xu, Sanjib K. Panda and Tong H. Lee is different from many books literature because it concentrates on the applications of iterative learning control in a wide range of industrial technologies.

At the heart of iterative learning control is the need to establish a contraction mapping so that the iterative learning converges, and to use an implementation architecture that can access the system error data trajectories for the control improvement updates. The way that updates are performed leads to several different architectures, for example, previous cycle learning, current cycle learning, previous and current cycle learning, and the practically important cascade iterative learning control method. This method places an iterative learning loop around a closed-loop system, thereby leaving the existing controller and its connections to the system intact; in this case it is the reference signal to the system that is updated. These fascinating developments are found in Chapter 2 and set the scene for the subsequent application chapters, of which there are eight. The range of the applications, which include mechatronics, electrical machines, process control, robotics, and PID controller design, together with details of the implementation solutions adopted are presented, making this a valuable complement to the existing theoretical literature, and a worthy new entry to the *Advances in Industrial Control* series.

Industrial Control Centre   *M.J. Grimble*
Glasgow   *M.A. Johnson*
Scotland, UK
2008

Preface

Iterative learning control (ILC) techniques have been successfully applied to solve a variety of real-life control-engineering problems, for example mechanical systems such as robotic manipulators, electrical systems such as electrical drives, chemical process systems such as batch reactors, as well as aerodynamic systems, bioengineering systems, and others. When such systems are operated repeatedly, iterative learning control can be used as a novel enabling technology to improve the system response significantly from trial to trial.

ILC is reputed for its promising and unique features: the structural simplicity, the perfect output tracking, almost model-independent design, and delay compensation. These highly desirable features make ILC a promising control alternative suitable for numerous real-time control tasks where a simple controller is required to achieve precise tracking in the presence of process uncertainties and delays.

In the past two decades, a great number of research studies focusing on ILC theory and performance analysis have been summarized and reported in dedicated volumes [1, 14, 20, 83, 153]. On the other hand, there is a lack of such a dedicated volume that can provide a wide spectrum of ILC designs, case studies and illustrative examples for real-time ILC applications In a sense, this book serves as a partial solution to meet the need in this specific area of control and applications. The ultimate objective of this book is to provide readers with the fundamental concepts, schematics, configurations and generic guidelines in ILC design and implementations, which are enhanced through a number of well-selected, representative, simple and easy-to-learn application examples.

In this book various key issues with regard to ILC design and implementations are addressed. In particular we discuss ILC design in the continuous-time domain and discrete-time domain, design in time and frequency domain, design with problem-specific performance objectives including both robustness and optimality, and design for parametric identification in open and closed-loop. The selected real-time implementations cover both linear and non-linear plants widely found in mechatronics, electrical drives, servo, and process control problems.

By virtue of the design and implementation nature, this book can be used as a reference for site engineers and research engineers who want to develop their own learning control algorithms to solve practical control problems. On the other hand, each control problem explored in this book is formulated systematically with the necessary analysis on the control-system properties and performance. Therefore, this book can also be used as a reference or textbook for a course at graduate level. Finally, we list open issues associated with the ILC design and analysis, and expect more academic researchers to look into and solve those challenging problems.

We would like to take the opportunity to thank our postgraduate students or research associates, Q.P. Hu, D.Q. Huang, W.Z. Qian, S.K. Sahoo, P. Srinivasan, Y. Tan, J. Xu, and H.W. Zhang, who made contributions to this book.

Singapore,   *Jian-Xin Xu*
August 2008   *Sanjib Kumar Panda*
                                                                    Tong Heng Lee

Contents

1 Introduction 1

2 Introduction to ILC: Concepts, Schematics, and Implementation 7
   2.1 ILC for Linear Systems 7
      2.1.1 Why ILC? 7
      2.1.2 Previous Cycle Learning 8
      2.1.3 Current Cycle Learning 10
      2.1.4 Previous and Current Cycle Learning 11
      2.1.5 Cascade ILC 12
      2.1.6 Incremental Cascade ILC 14
   2.2 ILC for Non-linear Systems 16
      2.2.1 Global Lipschitz Continuity Condition 17
      2.2.2 Identical Initialization Condition 19
   2.3 Implementation Issues 22
      2.3.1 Repetitive Control Tasks 22
      2.3.2 Robustness and Filter Design 23
      2.3.3 Sampled-data ILC 25
   2.4 Conclusion 27

3 Robust Optimal ILC Design for Precision Servo: Application to an XY Table 29
   3.1 Introduction 29
   3.2 Modelling and Optimal Indices 32
      3.2.1 Experimental Setup and Modelling 32
      3.2.2 Objective Functions for Sampled-data ILC Servomechanism 33
   3.3 Optimal PCL Design 35
   3.4 Optimal CCL Design 37
   3.5 Optimal PCCL Design 40
   3.6 Robust Optimal PCCL Design 42
   3.7 Conclusion 44

4 ILC for Precision Servo with Input Non-linearities: Application to a Piezo Actuator ... 47

4.1 Introduction ... 47
4.2 ILC with Input Deadzone ... 50
4.3 ILC with Input Saturation ... 53
4.4 ILC with Input Backlash ... 54
4.5 ILC Implementation on Piezoelectric Motor with Input Deadzone ... 55
    4.5.1 PI Control Performance ... 58
    4.5.2 ILC Performance ... 58
4.6 Conclusion ... 62

5 ILC for Process Temperature Control: Application to a Water-heating Plant ... 65
5.1 Introduction ... 65
5.2 Modelling the Water-heating Plant ... 67
5.3 Filter-based ILC ... 72
    5.3.1 The Schematic of Filter-based ILC ... 72
    5.3.2 Frequency-domain Convergence Analysis of Filter-based ILC ... 73
5.4 Temperature Control of the Water-heating Plant ... 76
    5.4.1 Experimental Setup ... 76
    5.4.2 Design of ILC Parameters M and γ ... 76
    5.4.3 Filter-based ILC Results for $\gamma = 0.5$ and $M = 100$ ... 78
    5.4.4 Profile Segmentation with Feedforward Initialization ... 78
    5.4.5 Initial Re-setting Condition ... 80
5.5 Conclusion ... 82
5.6 Appendix: The Physical Model of the Water-heating Plant ... 82

6 ILC with Robust Smith Compensator: Application to a Furnace Reactor ... 85
6.1 Introduction ... 85
6.2 System Description ... 86
6.3 ILC Algorithms with Smith Time-delay Compensator ... 88
6.4 ILC with Prior Knowledge of the Process ... 91
    6.4.1 ILC with Accurate Transfer Function ($P_0 = \hat{P}_0$) ... 91
    6.4.2 ILC with Known Upper Bound of the Time Delay ... 94
6.5 Illustrative Examples ... 95
    6.5.1 Simulation Studies ... 95
    6.5.2 Experiment of Temperature Control on a Batch Reactor ... 97
6.6 Conclusion ... 98

7 Plug-in ILC Design for Electrical Drives: Application to a PM Synchronous Motor ... 101
7.1 Introduction ... 101
7.2 PMSM Model ... 103

7.3   Analysis of Torque Pulsations 104
7.4   ILC Algorithms for PMSM 106
       7.4.1   ILC Controller Implemented in Time Domain 107
       7.4.2   ILC Controller Implemented in Frequency Domain 108
7.5   Implementation of Drive System 110
7.6   Experimental Results and Discussions 112
       7.6.1   Experimental Results 112
       7.6.2   Torque Pulsations Induced by the Load 116
7.7   Conclusion 120

8   **ILC for Electrical Drives: Application to a Switched Reluctance
    Motor** 121
8.1   Introduction 121
8.2   Review of Earlier Studies 124
8.3   Cascaded Torque Controller 124
       8.3.1   The TSF 125
       8.3.2   Proposed Torque to Current Conversion Scheme 126
       8.3.3   ILC-based Current Controller 128
       8.3.4   Analytical Torque Estimator 130
8.4   Experimental Validation of the Proposed Torque Controller 132
8.5   Conclusion 135

9   **Optimal Tuning of PID Controllers Using Iterative Learning
    Approach** 141
9.1   Introduction 141
9.2   Formulation of PID Auto-tuning Problem 144
       9.2.1   PID Auto-tuning 144
       9.2.2   Performance Requirements and Objective Functions 145
       9.2.3   A Second-order Example 145
9.3   Iterative Learning Approach 148
       9.3.1   Principal Idea of Iterative Learning 148
       9.3.2   Learning Gain Design Based on Gradient Information 150
       9.3.3   Iterative Searching Methods 153
9.4   Comparative Studies on Benchmark Examples 154
       9.4.1   Comparisons Between Objective Functions 155
       9.4.2   Comparisons Between ILT and Existing Iterative Tuning
               Methods 156
       9.4.3   Comparisons Between ILT and Existing Auto-tuning
               Methods 157
       9.4.4   Comparisons Between Search Methods 158
       9.4.5   ILT for Sampled-data Systems 160
9.5   Real-time Implementation 161
       9.5.1   Experimental Setup and Plant Modelling 161
       9.5.2   Application of ILT Method 162
       9.5.3   Experimental Results 163

9.6   Conclusion 163
9.7   Appendix 164
      9.7.1   Underdamped Case 164
      9.7.2   Overdamped Case 165
      9.7.3   Critical-damped Case 166

**10   Calibration of Micro-robot Inverse Kinematics Using Iterative
       Learning Approach** 169
      10.1   Introduction 169
      10.2   Basic Idea of Iterative Learning 171
      10.3   Formulation of Iterative Identifications 171
      10.4   Robustness Analysis with Calibration Error 175
      10.5   Example 176
             10.5.1   Estimation with Accurate Calibration Sample 177
             10.5.2   Estimation with Single Imperfect Factor in Calibration
                      Sample 178
             10.5.3   Estimation with Multiple Imperfect Factors in Calibration
                      Sample 179
      10.6   Conclusion 180

11   Conclusion 181

References 183

Index 191

Chapter 1
Introduction

In this book we present real-time iterative learning control (ILC) with successful applications to a number of processes widely encountered in industry. Many industrial control tasks are carried out repeatedly over a finite period. In such circumstances, control performance can be significantly enhanced by making full use of the process repeatability. Robotic manipulator control was the first successful application [4]. It was reported that the tracking error can be reduced to 1/1000 by 12 iterations with only simple changes of the reference signals [77].

Among various features, there are mainly four desirable features that make ILC an attractive control strategy in solving real-time control problems. The first desirable feature is the structural simplicity of ILC. In practical applications, a simple controller is always preferred, not only for the implementation cost but also for the control quality or reliability. The ILC mechanism can be as simple as an integral mechanism working iteratively. As such, ILC design becomes extremely simple, for instance, only one learning gain needs to be preset for single-input single-output systems. The pointwise integration of ILC can fully use process information such as the past tracking errors and past control signals over the entire operation period. ILC is a memory-based learning mechanism, and memory devices are extremely cheap with the present microprocessor technology.

The second desirable feature of ILC is the ability to achieve a perfect tracking both in the transient period and steady state with repeated learning. Up date, most control theories are still confined to stabilization as the best achievable result. Perfect tracking is tied in with the principle of internal model [34]. Instead of developing an explicit internal model, ILC develops the internal model implicitly through iteratively learning the control signals directly using a memory device. Therefore, ILC can learn arbitrary reference trajectories that may never enter steady state over the entire operation period.

The third desirable feature of ILC is its almost model-free nature in design and real-time execution. Unlike many control methods that require system model knowledge, ILC aims at the most difficult output tracking tasks, and meanwhile does not require any state knowledge except for a global Lipschitz condition. This is the most desirable feature for real-time control implementation, because seldom can we obtain an accurate plant model in practice, and in most circumstances the process modelling is a much more difficult and costly task in comparison with control.

The fourth desirable feature, and a unique feature, of ILC is the availability of non-causal signals for control compensation. By virtue of memory-based control updating, we can manipulate the control signals of previous trials, for instance using previous error signals at a time ahead of the current time instance. An immediate consequence is that we can easily compensate the process or sampling-time delay inherent in any feedback loops.

Despite the effectiveness of ILC witnessed in numerous applications, there is no dedicated book hitherto summarizing the recent advances in this active field. In the ILC literature, the first ILC book entitled "Iterative Learning Control for Deterministic Systems", authored by K.L. Moore and published by Springer-Verlag AIC series in 1993, concentrated on the ILC concept and several ILC algorithms [83]. The second ILC book entitled "Iterative Learning Control: Analysis, Design, Integration and Applications", co-edited by Z. Bien and J.-X. Xu and published by Kluwer Academic Press in 1998, introduced the latest ILC research results in theory and design up to 1997, and included three implementation cases in batching processing, arc welding and functional neuromuscular stimulation [14]. The third ILC book entitled "Iterative Learning Control: Convergence, Robustness and Applications", co-authored by Y.Q. Chen and C.Y. Wen and published by Springer-Verlag LNCIS series in 1999, focused on property analysis for discrete-time ILC algorithms [20]. The fourth ILC book entitled "Linear and Nonlinear Iterative Learning Control", co-authored by J.-X. Xu and Y. Tan and published by Springer-Verlag LNCIS series in 2003, mainly addressed theoretical problems in ILC [153]. The fifth ILC book entitled "Iterative Learning Control: Robustness and Monotonic Convergence in the Iteration Domain", co-authored by H.S. Ahn, Y.Q. Chen and K.L. Moore and published by Springer-Verlag Communications and Control Engineering series in 2007, focused also on a number of theoretical issues in ILC [1].

The purpose of this book is to provide a number of real-time ILC case studies that cover the ILC designs and applications in the fields of motion control and process control. As a consequence, this application-oriented book offers a complementary document to the existing theory and analysis oriented books in the ILC literature.

Before we state the contents of individual chapters of this book, it is worth briefly introducing recent advances in ILC research that directly relate to real-time ILC design and applications.

When implementing ILC, a critical issue is the ILC response along the iteration axis. It is well known that ILC aims to improve the transient control performance along the time domain, but it was also observed that ILC may have a poor response in the iteration domain [77, 151]. Many researchers have explored this issue and proposed specific ILC designs that can warrant a monotonic convergence in the supreme-norm along the iteration axis. In [91], the monotonic convergence property is made clear from the frequency domain, and some filter design is proposed. In [2], time-domain monotonic ILC design is proposed. However, the monotonic ILC problem still remains open for continuous-time linear systems and in general open for non-linear dynamic systems. Fortunately, it is also observed that many examples that show non-monotonic responses are often unstable or of lower stability margin in the time domain. In practical applications plants are either stable or made stable in the closed loop. As a result, the transient response in the iteration domain can be greatly improved.

Robustness and filter design are closely related in many control methods. As far as ILC is concerned, the robustness and filter design are two sides of a coin. Classical ILC performs an integral operation in the iteration domain, thus it could be sensitive to exogenous perturbations that appear in the iteration domain. A major source of iteration domain perturbations is the imperfect system repeatability. All real-time control devices have limited repeatability. For instance, the piezo motor applied to servo problems in Chap. 4, though having a precise position sensor resolution of 20 nanometers, can only provide a system repeatability at the precision level of 100 nanometers. Measurement noise and non-repeatable disturbances are also known to be another main source of perturbations. Various classes of filters were proposed to achieve robust ILC algorithms, mainly based on the prior knowledge of the plant model and the spectrum or the stochastic characteristics of perturbations. An early study [77] provided comprehensive considerations on filter design for continuous-time ILC. Recently, a good summary has been documented in a survey article [15] where discrete-time ILC is concerned. In the next chapter, we will briefly revisit this critical issue on filter design. It still remains an open problem on robust ILC for non-linear dynamic processes, because frequency-domain analysis is not directly applicable. An alternative approach for non-linear dynamics is to implement ILC in the discrete frequency domain using Fourier transforms [149]. The rationale lies in the fact that the repeated control profiles consist of discrete frequency components only. It is adequate to learn the frequencies of interests, usually at the low-frequency band. The attempt to learn frequencies near the Nyquist frequency or the actuator bandwidth is impossible. Further, any two consecutive iterations are essentially disconnected executions, therefore the time-consuming frequency analysis and processing can be conducted in between two iterations.

The majority of industrial control problems concentrate on motion control and process-control problems. Among numerous motion control problems, the first ILC application, and also the first ILC paper, was on robotic control [4]. Up to now, robotics is still a good testbed for the verification of ILC algorithms. In [90] each link of the robotic arm can be modelled simply as a first-order system. ILC was able to learn and compensate the modelling uncertainties. In [131], adaptive ILC algorithms were developed. It is worth highlighting that [131] gave the first real-time ILC application in which the iterative learning controller was designed using the Lyapunov method.

Many of the application-oriented ILC works focused on servo problems and electrical drives. The popularity of ILC research in this area is owing to the fact that most actuators nowadays used in motion-control systems or mechatronics are servo for the position control and electrical drives for the velocity control, comparing with other actuators such as hydraulic or pneumatic actuators [18]. With ILC designed either in the time domain or in the frequency domains, these applications ubiquitously exhibit significant improvements in performance when the control task repeats. In this book we demonstrate various servo and electrical drive designs and applications through five dedicated chapters. In particular we show a promising advantage of ILC in Chap. 4, *i.e.* ILC can adapt to the underlying non-linear and non-smooth factors presented in most actuators, such as deadzone, saturation and backlash.

Process control comprises a large class of industrial control tasks, including temperature-trajectory tracking, pressure control, level control, concentration control, *etc.* ILC applications to process control problems can be seen from a number of published reports. Two representative ILC applications are wafer industry and chemical reactors. In [161], an optimal ILC was applied to wafer-temperature control in rapid thermal processing problems. In [81], ILC was applied to an exothermic semi-batch chemical reactor. Comparing with motion control, ILC applications to process control are limited and more effort is needed.

In real-time industrial control problems, proportional-integral-derivative (PID) and model predictive control (MPC) are two predominant and matured control technologies that constitute more than 90% of feedback loops. A number of PID-type ILC and model-predictive ILC algorithms have been exploited. PID-type ILC can be constructed in two ways. The first way is to use PID errors to update the current control inputs. In fact the two most classical ILC algorithms, P-type and D-type ILC, use proportional and derivative signals, respectively, in learning updating. We will elaborate this design approach in Chaps. 2 and 3. The second way is to add an ILC mechanism on to an existing PID control loop in a modular approach. Two examples are shown in Chap. 7 and [45]. PID-type ILC inherits the most desirable features of PID: simple and almost model free. The add-on approach is widely adopted in real-time applications owing to the integration of closed-loop feedback in the time domain and learning in the iteration domain. Model-predictive ILC, on the other hand, provides a systematic approach to design a controller in an optimal sense associated with a selected objective function. Model-predictive ILC has been studied [3, 12, 72]. Using a quadratic objective function, the controller can be designed based on the nominal system. The modelling errors and exogenous disturbances, as far as repeatable around the reference trajectory, can be compensated by ILC.

The outline of this book is as follows.

Chapter 2 introduces the fundamental ILC schematics, including the previous cycle learning (PCL), current cycle learning (CCL), previous and current cycle learning (PCCL), embedded structure, cascade structure, as well as learning convergence conditions associated with individual ILC schemes. This chapter serves as a rudimentary introduction to ILC so as to make the book self-contained for most readers such as control engineers and graduate students who may not be familiar with ILC theory.

Chapter 3 presents a robust optimal ILC design method based on 1) the plant nominal model, which is linearized in the discrete-time, and 2) the range of process modelling uncertainties. Through minimizing some objective functions defined in the frequency domain, ILC algorithms for an XY table are designed and verified experimentally.

Chapter 4 demonstrates that the simplest ILC designed in the discrete-time domain can effectively compensate the non-smooth and non-linear factors in the system input, for instance deadzone, saturation and backlash. Subsequently, the ILC is applied to a piezo motor and achieved accurate tracking performance experimentally in the presence of a state-dependent unknown deadzone.

Chapter 5 provides a detailed frequency-domain design for a heat-exchange process that is linearized in continuous time. The classical Bode-plot approach can be directly applied for the ILC and filter design, and satisfactory results are obtained experimentally.

Chapter 6 describes how the classical Smith prediction technique can be integrated with ILC to deal with a linear continuous-time plant with large dead-time and model mismatching. The combined ILC scheme is applied to the temperature control of a furnace that produces new materials.

Chapter 7 studies the speed control of a permanent magnetic synchronous motor (PMSM) where ILC is used to minimize the undesired speed ripples. A plug-in ILC is designed in discrete time, and implemented in both the time domain and the frequency domain using fast Fourier transformation. Experimental results show that, though non-linear in nature, the frequency-domain ILC can perform robustly to measurement noise and other non-repeatable perturbations.

Chapter 8 addresses the ILC design for and application to a highly non-linear switched reluctance motor (SRM), which is particularly non-linear both in system inputs and states. The ILC block is added on to the existing control loop to improve the control performance, and is validated through experiments on a real setup.

Chapter 9 details an iterative learning algorithm for optimal tuning of PID parameters. An objective function is employed to quantify the time-domain transient response, such as the overshoot, settling time, *etc.* By searching the PID parameters iteratively, the objective function is minimized, and subsequent control performance is improved. Experiments on the level control of a coupled-tank system verifies the effectiveness of the iterative-learning-based PID tuning.

Chapter 10 develops a specific iterative-learning-based method for system identification and calibration. Two highly desirable features associated with this method are 1) ability to identify or calibrate process parameters using only a few or even a single measurement sample, and 2) ability to deal with processes non-linear in the parametric space. The validity of the method is verified through the kinematics and inverse kinematics identification of a multi-link closed-chain micro-robotic manipulator.

Chapter 11 concludes the book and points out several future research directions closely related to ILC design and implementation.

Chapter 2
Introduction to ILC: Concepts, Schematics, and Implementation

Abstract In this chapter we review several important concepts, basic schematics, and implementation issues associated with real-time ILC. We first introduce five basic configurations of ILC for linear processes, including the previous cycle learning, current cycle learning, previous and current cycle learning, cascade learning, and incremental cascade learning. Next, we focus on ILC for non-linear processes, make clear two conditions intrinsic to ILC applications – the global Lipschitz continuity condition and identical initialization condition, and explore possible extensions. Finally we discuss three practical issues encountered in real-time ILC implementation – repetitive control tasks, robustness and filter design, as well as the sampling effect.

2.1 ILC for Linear Systems

In this section the concepts and schematics of ILC will be briefly reviewed for linear systems.

2.1.1 Why ILC?

Consider a control task that requires the perfect tracking of a pre-specified reference trajectory, for example moving and fixing parts in an assembly line, or temperature control of a batch reactor in the pharmaceutical industry. The common features of this class of control problems are 1) the task must be finished in a finite duration ranging from milliseconds to days, 2) the reference trajectory must be strictly followed from the very beginning of the execution, 3) the task is repeated from trial to trial, from batch to batch, from run to run, or in general from iteration to iteration, under the same conditions.

We face a new class of control tasks: perfect tracking in a finite interval under a repeatable control environment, where the repeatable control environment stands for the repeatability of the process under the given control task that repeats and in the presence of repeated disturbances.

Most existing control methods including adaptive or robust control, may not be suitable for such a class of tasks because of two reasons. First, these control methods are characterized by the asymptotic convergence, thus it is difficult to guarantee a perfect tracking even if the initial discrepancy is zero. Second and more importantly, those control methods are not able to "learn" from previous task execution that may succeed or fail. Without learning, a control system can only produce the same performance without improvement even if the task is repeatedly executed. ILC was proposed to meet this kind of control requirement [4, 14, 83]. The idea of ILC is straightforward: using control information of the preceding execution to improve the present execution. This is realized through memory-based learning.

Iterative learning controllers can be constructed in many different ways. In this section we demonstrate five representative and most commonly used ILC configurations. In general, ILC structures can be classified into two major categories: embedded and cascaded. In the following, the first three belong to the embedded structure, and the next two belong to the cascade structure.

2.1.2 Previous Cycle Learning

The configuration of a previous cycle learning (PCL) scheme is shown in Fig. 2.1. Here, the subscript i denotes the ith iteration. Notations $Y_{r,i}(s)$, $Y_i(s)$, $U_i(s)$ and $E_i(s)$ denote the reference signals, output signals, control signals, and error signals, respectively at the ith iteration. $P(s)$ and $C_l(s)$ denote the transfer functions of the plant and the feedforward compensator, respectively. When the system performs the same control task, $Y_{r,i+1}(s) = Y_{r,i}(s)$. The MEM labelled with Y, Y_r and U are memory arrays storing system signals of the current cycle, $i.e.$ $(i+1)$th iteration, which will be used in the next learning cycle (iteration).

Assume that the control task is repeated for all iterations, that is, $Y_r = Y_{r,i} = Y_{r,i+1}$. According to the PCL configuration shown in Fig. 2.1,

$$Y_i = PU_i$$
$$E_i = Y_r - Y_i$$
$$U_{i+1} = U_i + C_l E_i. \tag{2.1}$$

Equation 2.1 is the PCL updating law. It is called previous cycle learning simply because only the previous cycle control signals U_i and previous error signals E_i are used to form the current cycle control input U_{i+1}. It is an open-loop control in the time domain, but a closed-loop control in the iteration domain.

The learning convergence condition for PCL can be derived as,

$$E_{i+1} = Y_r - Y_{i+1}$$
$$= Y_r - PU_{i+1}$$

$$= Y_r - P(U_i + C_l E_i)$$
$$= Y_r - PU_i + PC_l E_i$$
$$= (1 - PC_l)E_i$$
$$\Rightarrow \qquad \frac{E_{i+1}}{E_i} = 1 - PC_l$$
$$\Rightarrow \qquad \|\frac{E_{i+1}}{E_i}\| = \|1 - PC_l\| \le \gamma < 1, \qquad (2.2)$$

where the norm $\| \cdot \|$ is the infinity norm for all frequencies $\omega \in \Omega$, $\Omega = [\omega_a, \omega_b]$, $\omega_b > \omega_a \ge 0$. Ω denotes the frequency band of interest, or the frequency band that matches the bandwidth of a controller. Clearly, as far as the tracking error signals of the 1st iteration, E_0, is finite, then $\|E_i\| \le \gamma^i \|E_0\| \to 0$ as $i \to \infty$. It can also be seen that, for a process P, an appropriate choice of C_l will make a smaller γ and hence expedite the learning process.

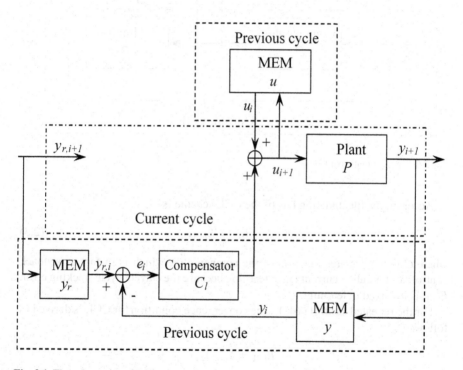

Fig. 2.1 The schematics of PCL

2.1.3 Current Cycle Learning

Due to the open-loop nature in the current cycle, PCL could be sensitive to small and non-repeatable perturbations. This can be improved by a feedback-based learning if the loop can be closed appropriately in the time domain, leading to the current cycle control (CCL). The configuration of the CCL scheme is shown in Fig. 2.2.

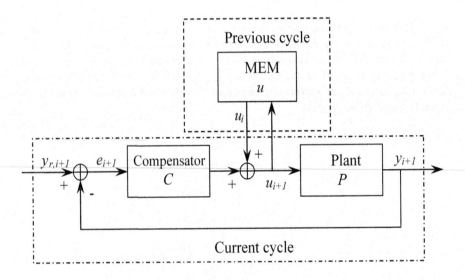

Fig. 2.2 The schematics of CCL

Accordingly, the updating law of the CCL scheme is

$$U_{i+1} = U_i + CE_{i+1}, \tag{2.3}$$

where C is the transfer function of the compensator, which is in fact a feedback controller. It is called current cycle learning because the current cycle tracking error, E_{i+1}, is involved in learning.

For the repeated control task Y_r, the convergence condition for CCL is derived as follows

$$
\begin{aligned}
E_{i+1} &= Y_r - Y_{i+1} \\
&= Y_r - PU_{i+1} \\
&= Y_r - P(U_i + CE_{i+1}) \\
&= Y_r - PU_i - PCE_{i+1} \\
(1+PC)E_{i+1} &= E_i \\
\Rightarrow \quad \frac{E_{i+1}}{E_i} &= \frac{1}{1+PC}
\end{aligned}
$$

$$\Rightarrow \qquad \left\| \frac{E_{i+1}}{E_i} \right\| = \left\| \frac{1}{1+PC} \right\| \le \gamma < 1. \qquad (2.4)$$

It can be seen that PCL and CCL are functioning in a complementary manner. PCL requires the factor $\|1 - PC_l\|$ to be below 1 for all frequencies within the band Ω, and often leads to a low gain C_l. CCL requires the factor $\|1 + PC\|$ to be above 1 for all frequencies within the band Ω, and often a high gain C is preferred as far as the closed-loop stability is guaranteed. Note that the memory arrays required for CCL are half of the PCL, because MEM for E_i is not required. Memory size is an important factor when implementing an ILC scheme.

2.1.4 Previous and Current Cycle Learning

Generally speaking, it may be difficult to find a suitable C_l such that (2.2) is satisfied, i.e. $\|1 - PC_l\|$ is strictly less than 1 for all frequencies in Ω. Likewise, it may be a difficult job to find a suitable C such that the denominator in (2.4) is strictly larger than 1 for all frequencies in Ω. There is only 1 degree of freedom (DOF) in the controller design for both schemes. By pairing PCL and CCL together to form the previous and current cycle learning (PCCL), there is a possibility that the learning convergence will be improved. This can be achieved if the convergence conditions (2.2) and (2.4) can complement each other at frequencies where one convergence condition is violated. There are several ways to pair the PCL and CCL, and a possible combination is shown in Fig. 2.3.

For the repeated control task Y_r, it can be easily derived that the convergence condition is

$$\frac{\|1 - PC_l\|}{\|1 + PC\|},$$

that is, the multiplication of both conditions of PCL and CCL. To demonstrate the advantage of the 2 DOF design with PCCL, an example is shown in Fig. 2.4, where

$$1 - PC_l = \frac{s^2 - 3}{3s^2 + 3s + 5}$$

and

$$1/(1 + PC) = \frac{s^2 + 20s + 80}{2s^2 + 10s + 130}.$$

Note that neither the PCL condition (2.2) nor the CCL condition (2.4) holds for all frequencies within the band $\Omega = [10^{-2}, 10^2]$ rad/s, but the convergence condition with the integrated PCCL, which is the superposition of PCL and CCL in the Bode plot, does hold for all frequencies.

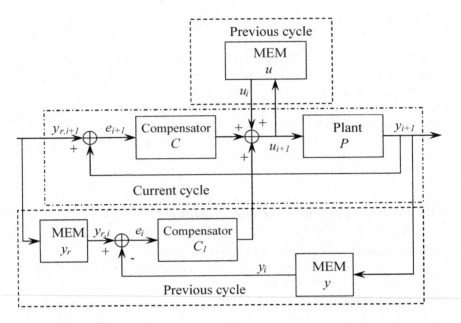

Fig. 2.3 The schematics of a PCCL

2.1.5 Cascade ILC

In the previous three ILC schemes, one may observe that the control system has been redesigned, with a new feedforward component added to the system input channel. From the configurations shown in Figs. 2.1, 2.2 and 2.3, a new control block is embedded into the control loop. Such an embedded structure is the common structure for most existing real-time ILC schemes. Hence, when an ILC mechanism is to be incorporated into an existing control system, either the core execution program needs to be re-written or the micro-controller chip needs to be replaced. In many real applications, such a re-configuration of a commercial controller is not acceptable due to the cost, security and intellectual-property constraints. For instance, a rapid thermo-processing device in the wafer industry costs millions of dollars, and the only tunable part is a number of setpoints. In such circumstance, the cascade learning method is suitable as it modifies only the reference trajectory iteratively to improve the control performance.

The schematics of such an ILC is demonstrated in Fig. 2.5. It can be seen that the ILC block is "cascaded" to the existing control loop or performed as an outer feedback loop in the iteration domain. The ILC with the cascade structure will use the modified reference signals and the actual system output of the previous cycle to generate the new reference signals for the current cycle. Owing to the cascade structure, the ILC need not be embedded into the existing control loop, thus avoiding any re-configuration of the system hardware or the core execution program. What is needed is essentially some re-programming of reference signals, which can be easily carried out in real applications.

According to Fig. 2.5, we have

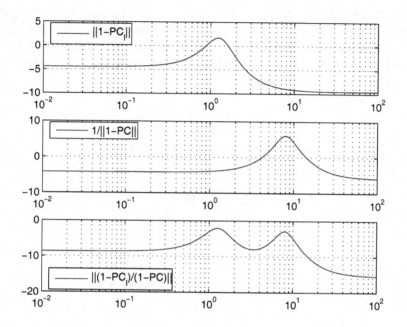

Fig. 2.4 The complementary role of PCL and CCL

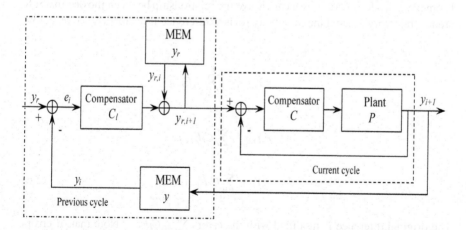

Fig. 2.5 The schematics of a cascade ILC structure

$$Y_i = GY_{r,i}$$
$$E_i = Y_r - Y_i$$
$$Y_{r,i+1} = Y_{r,i} + C_l E_i \tag{2.5}$$
$$Y_{r,0} = Y_r,$$

where $G = PC/(1 + PC)$ denotes the closed-loop transfer function; Y_r is the original reference repeated over a fixed operation period; $Y_{r,i}$ is the reference signal modified via learning for the inner current cycle control loop. According to the learning control law (2.5), the convergence condition for the cascade ILC can be derived as

$$
\begin{aligned}
E_{i+1} &= Y_r - Y_{i+1} \\
&= Y_r - GY_{r,i+1} \\
&= Y_r - G(Y_{r,i} + C_l E_i) \\
&= Y_r - GY_{r,i} - GC_l E_i \\
&= (1 - GC_l)E_i
\end{aligned}
$$

$$\Rightarrow \qquad \frac{E_{i+1}}{E_i} = 1 - GC_l$$

$$\Rightarrow \qquad \|\frac{E_{i+1}}{E_i}\| = \|1 - \frac{PCC_l}{1+PC}\| \le \gamma < 1.$$

In most cases the cascade ILC is of PCL type, because set points, once selected, cannot be changed in the midst of a real-time operation process.

2.1.6 Incremental Cascade ILC

From the cascade ILC (2.5), we can derive the relationship between the original reference trajectory, Y_r, and the iteratively revised reference trajectory, $Y_{r,i+1}$, as below

$$
\begin{aligned}
Y_{r,i+1} &= Y_{r,i} + C_l E_i \\
&= Y_{r,i-1} + \sum_{j=0}^{1} C_l E_{i-j} \\
&= \cdots = \\
&= Y_{r,0} + \sum_{j=0}^{i} C_l E_{i-j} \\
&= Y_r + \sum_{j=0}^{i} C_l E_{i-j}. \tag{2.6}
\end{aligned}
$$

The original reference is modified with the errors $\sum_{j=0}^{i} C_l E_{i-j}$. Note that, if errors $E_0, \cdots, E_i$ are zero, $Y_{r,i+1} = Y_r$, that is, the original reference shall be used. In other words, if the current cycle feedback loop is able to achieve perfect tracking with regard to the original reference Y_r, then the cascade ILC shall retain the original reference trajectory.

A potential problem associated with the cascade ILC in real-time applications is the filtering. In practical control processes with ILC, filters are widely applied and are imperative in general, due to the existence of non-repeatable perturbations or measurement noise. We will address the filtering issue in a later part of this chapter. A common practice in ILC applications is to design a low-pass filter or a band-pass filter, $F[\cdot]$, for the preceding control signals stored in memory, namely

$$U_{i+1} = F[U_i] + C_l E_i$$

for the embedded ILC, or

$$Y_{r,i+1} = F[Y_{r,i}] + C_l E_i$$

for the cascade ILC. The cascade ILC with filtering becomes

$$Y_{r,i+1} = F[Y_{r,i}] + C_l E_i$$

$$= F^2[Y_{r,i-1}] + \sum_{j=0}^{1} F^j[C_l E_{i-j}]$$

$$= \cdots$$

$$= F^i[Y_{r,0}] + \sum_{j=0}^{i} F^j[C_l E_{i-j}]$$

$$= F^i[Y_r] + \sum_{j=0}^{i} F^j[C_l E_{i-j}], \tag{2.7}$$

where $F^0[x] = x$. It is clear that, even if errors $E_0, \cdots, E_i$ are all zero, $Y_{r,i+1} = F^i[Y_r] \neq Y_r$, that is, the original reference has been changed. In other words, while the feedback loop is able to achieve perfect tracking with regard to the original reference Y_r, the cascade ILC is unable to retain the original reference trajectory, instead it is only able to provide a filtered one. As a result, Y_r will be distorted by $F^i[\cdot]$ with additional magnitude change and phase lag, usually the larger the number of filtering times i, the larger the distortion. Clearly the original tracking task cannot be achieved in such a circumstance.

A solution to this problem is to introduce an incremental structure in the cascade ILC, as shown in Fig. 2.6. The learning control law is

$$Y_{r,i+1} = Y_r + \Delta Y_{r,i+1}$$
$$\Delta Y_{r,i+1} = \Delta Y_{r,i} + C_l E_i$$
$$\Delta Y_{r,0} = 0. \tag{2.8}$$

It is easy to derive that incremental learning is the same as the previous cascade learning (2.6),

$$Y_{r,i+1} = Y_r + \Delta Y_{r,i+1}$$

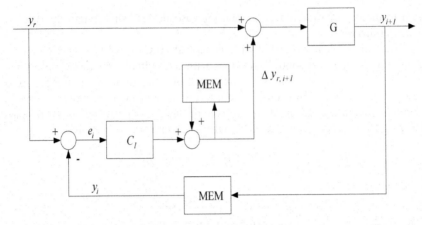

Fig. 2.6 The schematics of an incremental cascade ILC structure

$$= Y_r + \Delta Y_{r,i} + C_l E_i$$

$$= \cdots$$

$$= Y_r + \sum_{j=0}^{i} C_l E_{i-j}. \tag{2.9}$$

Now, adding a filter such that $\Delta Y_{r,i+1} = F[\Delta Y_{r,i}] + C_l E_i$, we can derive

$$
\begin{aligned}
Y_{r,i+1} &= Y_r + \Delta Y_{r,i+1} \\
&= Y_r + F[\Delta Y_{r,i}] + C_l E_i \\
&= Y_r + F^2[\Delta Y_{r,i-1}] + \sum_{j=0}^{1} F^j[C_l E_{i-j}] \\
&= \cdots \\
&= Y_r + \sum_{j=0}^{i} F^j[C_l E_{i-j}].
\end{aligned}
\tag{2.10}
$$

It can be seen that the filtering applies to the past tracking errors. If errors $E_0, \cdots, E_i$ are all zero, $Y_{r,i+1} = Y_r$, that is, the original reference will not be changed.

2.2 ILC for Non-linear Systems

In practical applications, most plants are characterized by non-linear dynamics. In this section, we investigate the applicability of linear ILC schemes for non-linear dynamical processes. Consider the following non-linear plant

$$\dot{\mathbf{x}}(t) = \mathbf{f}(\mathbf{x}(t), u(t)) \qquad \mathbf{x}(0) = \mathbf{x}_0$$

$$y(t) = g(\mathbf{x}(t), u(t)), \tag{2.11}$$

where $t \in [0, T]$, $\mathbf{x}(t) \in \mathscr{R}^n$, $y(t) \in \mathscr{R}$ and $u \in \mathscr{R}$, $\mathbf{f}(\cdot)$ is a smooth vector field, and $g(\cdot)$ is a smooth function.

Under the repeatability of the control environment, the control objective for ILC is to design a sequence of appropriate control inputs $u_i(t)$ such that the system output $y_i(t)$ approaches the target trajectory $y_d(t)$, $\forall t \in [0, T]$. In Arimoto's first ILC article [4] a typical yet the simplest first-order linear iterative learning control scheme is proposed

$$u_{i+1}(t) = u_i(t) + \beta e_i(t), \tag{2.12}$$

where $e_i = y_r - y_i$, and β is a constant learning gain. The initial control profile $u_0(t)$ $t \in [0, T]$ is either set to zero or initialized appropriately by some control mechanism. The convergence condition is determined by the relation

$$|u_{i+1}|_\lambda \le |1 - \beta g_u| |u_i|_\lambda$$
$$|1 - \beta g_u| = \gamma < 1, \tag{2.13}$$

where $g_u(\mathbf{x}, u, t) = \frac{\partial g}{\partial u}$, and $|\cdot|_\lambda$ is the time-weighted norm defined as $\max_{t \in [0,T]} e^{-\lambda t} |\cdot|$. The learnability condition is that the system gain $g_u \in [\alpha_1, \alpha_2]$, either $\alpha_1 > 0$ or $\alpha_2 < 0$. By choosing a proper learning gain β, the convergence condition (2.13) can be fulfilled if the interval $[\alpha_1, \alpha_2]$ is known *a priori*.

There are two conditions indispensable for linear ILC to be applicable to the non-linear process (2.11): the global Lipschitz continuity condition (GLC) and identical initialization condition (perfect resetting).

2.2.1 Global Lipschitz Continuity Condition

The GLC condition states that $\mathbf{f}(\mathbf{x}, u, t)$ and $g(\mathbf{x}, u, t)$ are global Lipschitz continuous with respect to the arguments. For $\mathbf{x}_1, \mathbf{x}_2 \in \mathscr{R}^n$, and $u_1, u_2 \in \mathscr{R}$, there exist constants L_f, L_g such that

$$\|\mathbf{f}(\mathbf{x}_1, u_1) - \mathbf{f}(\mathbf{x}_2, u_2)\| \le L_f(\|\mathbf{x}_1 - \mathbf{x}_2\| + |u_1 - u_2|)$$

and

$$\|\mathbf{g}(\mathbf{x}_1, u_1) - \mathbf{g}(\mathbf{x}_2, u_2)\| \le L_g(\|\mathbf{x}_1 - \mathbf{x}_2\| + |u_1 - u_2|).$$

Clearly, a linear plant is a special case of GLC functions. For instance, $\mathbf{f} = A\mathbf{x} + \mathbf{b}u$ has two Lipschitz constants denoted by an induced matrix norm $\|A\|$ and a vector norm $\|\mathbf{b}\|$.

In practical applications, it might not be easy to directly check whether a non-linear function $\mathbf{f}(\mathbf{x})$ satisfies the GLC condition. A more convenient way is to derive the Jacobian of the non-linear function

$$\frac{\partial \mathbf{f}}{\partial \mathbf{x}}.$$

From the continuity of $\mathbf{f}$, applying the mean value theorem leads to

$$\mathbf{f}(\mathbf{x}_1) - \mathbf{f}(\mathbf{x}_2) = \frac{\partial \mathbf{f}}{\partial \mathbf{x}}\Big|_{\mathbf{x}=\xi} (\mathbf{x}_1 - \mathbf{x}_2)$$

where $\xi \in [\mathbf{x}_1, \mathbf{x}_2]$. Hence we obtain a condition similar to GLC

$$\|\mathbf{f}(\mathbf{x}_1) - \mathbf{f}(\mathbf{x}_2)\| \leq \|\frac{\partial \mathbf{f}}{\partial \mathbf{x}}\|\|\mathbf{x}_1 - \mathbf{x}_2\|$$

and the Lipschitz constant can be evaluated by

$$\max_{\xi \in [\mathbf{x}_1, \mathbf{x}_2]} \|\frac{\partial \mathbf{f}}{\partial \mathbf{x}}\Big|_{\mathbf{x}=\xi}\|.$$

In the above derivation, the variables ξ are confined to a region specified by $[\mathbf{x}_1, \mathbf{x}_2]$. A sufficient condition for a non-linear function to be globally Lipschitz continuous is that its Jacobian is globally bounded, that is, uniformly bounded for any $\xi \in \mathscr{R}^n$. In fact, the quantity g_u in the learning convergence condition (2.13) is exactly the Jacobian.

Many engineering systems do not meet the GLC condition. For instance, as we will discuss later in this book, the current–torque (input–output) relationship of a switched reluctance motor contains a non-linear factor xe^{ax}, where x is the stator current and a is a bounded parameter. Its Jacobian is $(1+ax)e^{ax}$, which is not uniformly bounded for any $x \in \mathscr{R}$. Nevertheless, the physical variables are often limited. For example, all electrical drives are equipped with current limiters to confine the maximum current to flow through the stator winding. As such, the Jacobian is always finite, subsequetly the GLC condition is satisfied for the switched reluctant motor under any operating conditions.

In practical applications we also encounter such systems

$$\dot{\mathbf{x}}(t) = \mathbf{f}(\mathbf{x}(t), u(t)) \qquad \mathbf{x}(0) = \mathbf{x}_0$$
$$y(t) = g(\mathbf{x}(t)), \tag{2.14}$$

where the output y is not a function of the input u. In such circumstances, the preceding simple learning control law (2.12), usually called the P-type ILC law, does not work. Let us choose $\dot{e}$ as the new output variable, then the new input–output relationship is

$$\dot{e}(t) = \frac{\partial g(\mathbf{x}(t))}{\partial \mathbf{x}} \mathbf{f}(\mathbf{x}(t), u(t)). \tag{2.15}$$

A D-type ILC law is constructed accordingly

$$u_{i+1}(t) = u_i(t) + \beta \dot{e}_i(t). \tag{2.16}$$

The convergence condition is determined by the relationship

$$|u_{i+1}|_\lambda \leq |1 - \beta J||u_i|_\lambda$$
$$|1 - \beta J| = \gamma < 1, \tag{2.17}$$

where

$$J = \left(\frac{\partial g}{\partial \mathbf{x}}\right)^T \frac{\partial \mathbf{f}}{\partial u}$$

is the Jacobian of the new input–output relationship.

Note that what is ensured by the convergence condition (2.17) is the asymptotic convergence of the error derivative $\Delta \dot{y}_i(t) \to 0$ as $i \to \infty$. By virtue of the identical initial condition $e_i(0)$, as we will discuss next, the property $\dot{e}_i(t) = 0$ implies $e_i(t) = 0$ for any $t \in [0, T]$.

2.2.2 Identical Initialization Condition

Initial conditions, or initial resetting conditions, play a fundamental role in all kinds of iterative learning control methods [97]. ILC tries to make a "perfect" tracking in a finite time interval, whereas other control methods aim at an asymptotical convergence in an infinite time interval.

In practice, a control task must be finished in a finite time interval, for example the track following control of a hard disk drive with the duration in milliseconds, or temperature control of a batch reactor with the duration in hours or days. Note that these control problems require perfect tracking from the beginning. Consequently many existing control methods with the asymptotical convergence property are not suitable. A perfect tracking from the beginning demands the perfect initial condition, that is, the identical initialization condition (*i.i.c.*). In fact, in order to fulfill such a control task, other control methods such as adaptive control or robust control will also need the same condition.

Why do most existing control methodologies not require this condition? This is because they allow the control task to be achieved asymptotically in an infinite time interval, hence the initial discrepancy is not a problem. ILC cannot enjoy such an asymptotical convergence in the time domain because the task must be finished in a finite time interval, and requires perfect tracking over the entire transient period including the initial one.

In many practical control problems, the *i.i.c.* can be easily satisfied. For instance, most mechatronics systems have homing or re-setting mechanisms. Thus, we know where the process starts from. A batch reactor in a bio-chemical process always starts from a known room temperature that is under regulation.

There are three structural initialization conditions [153] used in ILC: two are state related and one is output related, mathematically expressed as (i) $\mathbf{x}_i(0) = \mathbf{x}_0(0)$, (ii) $\mathbf{x}_i(0) = \mathbf{x}_r(0)$, (iii) $y_i(0) = y_r(0)$. Note that the second condition implies the first condition, hence is more restrictive. The third condition is required in particular by the D-type ILC as previously discussed. For strictly repeatable systems like (2.14), the second condition also implies the third condition.

In some practical applications, however, the perfect initial re-setting may not be obtainable. Five different initial conditions were summarized in [158],
(a) identical initial condition;
(b) progressive initial condition, *i.e.* the sequence of initial errors belong to l^2;
(c) fixed initial shift;
(d) random initial condition within a bound;
(e) alignment condition, *i.e.* the end state of the preceding iteration becomes the initial state of the current iteration.

The preceding three structural initialization conditions (i), (ii), (iii) were special cases of the condition (a). We face an important issue in ILC applications: the robustness under conditions (b), (c), (d) and (e). Most relevant studies done along this line were discussed in the literature survey of [158]. The simplest way is to revise the ILC updating law with a forgetting factor. For example, the learning law (2.12) becomes

$$u_{i+1}(t) = \alpha u_i(t) + \beta w_i(t), \tag{2.18}$$

where $\alpha \in (0, 1]$ is a forgetting factor, and $w_i(t)$ stands for one of the quantities $e_i(t)$, $e_{i+1}(t)$, or $\dot{e}_i(t)$. The effect of α is to prevent the influence of an initial discrepancy from going unbounded. Let us show this robustness property. Assume that the initial error, $w_i(0)$, is randomly distributed within a bound of ε, *i.e.* the condition (d) that is considered the most difficult yet most practical scenario. Substituting (2.18) to itself repeatedly with descending i we have

$$
\begin{aligned}
u_{i+1}(0) &= \alpha u_i(0) + \beta w_i(0) \\
&= \alpha^2 u_{i-1}(0) + \alpha \beta w_{i-1}(0) + \beta w_i(0) \\
&= \cdots \\
&= \alpha^{i+1} u_0(0) + \beta \sum_{j=0}^{i} \alpha^{i-j} w_j(0).
\end{aligned}
\tag{2.19}
$$

Initial control $u_0(0)$ is always finite in practical problems, hence $\alpha^{i+1} u_0(0) \to 0$ when $i \to \infty$. Therefore, as $i \to \infty$,

$$|u_{i+1}(0)| \le \beta \sum_{j=0}^{i} \alpha^{i-j} |w_j(0)| \tag{2.20}$$

$$\le \beta \varepsilon \frac{1 - \alpha^{i+1}}{1 - \alpha}, \tag{2.21}$$

which is obviously finite as far as $\alpha \in (0, 1)$. In real-time ILC implementations, some prior knowledge is needed in choosing an appropriate value for α. A recommended range is $[0.95, 0.99]$. However, from (2.21) the error bound could be 20–100 times higher than ε when choosing $\alpha \in [0.95, 0.99]$. A lower α can reduce the error bound, but meanwhile the learning effect is inevitably weakened. To avoid such a tradeoff, we need to look into initial conditions (b), (c), (d) in more detail.

The issue concerned with the condition (b) is whether the accumulated influence of the initial discrepancy will vanish if the initial error is itself vanishing gradually, *i.e.* under the progressive condition (b). Assume that $w_i(0)$ is gradually vanishing with a geometric rate, hence its upper bound can be expressed by $\gamma_1^i \varepsilon$ with $\gamma_1 \in (0, 1)$. Denote $\gamma_2 = \max\{\alpha, \gamma_1\}$. Clearly $\gamma_2 \in (0, 1)$. According to the preceding derivation,

$$
\begin{aligned}
|u_{i+1}(0)| &\leq \beta \sum_{j=0}^{i} \alpha^{i-j} \gamma_1^j \varepsilon \\
&\leq \beta \varepsilon \sum_{j=0}^{i} \gamma_2^{i-j} \gamma_2^j \\
&= (i+1) \gamma_2^i \beta \varepsilon
\end{aligned}
\tag{2.22}
$$

which goes to zero when $i \to \infty$. In such circumstances, we can choose α sufficiently close to 1 without sacrificing the ultimate error bound.

The condition (c) indicates the presence of a fixed initial shift, for example an XY table homing position is not the starting point of the task, or the room temperature is not the starting temperature of a reactor. A set of initial state controllers, such as proportional integral (PI) controllers, can be used to bring the initial states to the desired values, then the tracking process starts. This way is practically easy and effective, as it is a point-to-point or setpoint control problem that is much easier than tracking control problems. If the initial state control is not possible, an alternative way is to rectify the target trajectory at the initial phase. For D-type ILC, we can set $\mathbf{y}_r(0) = \mathbf{y}_i(0)$, because the output is accessible [123]. On the other hand, *i.i.c* is still open when the initial states $\mathbf{x}_i(0)$ are not accessible.

The scenario (d) is most frequently encountered in practice. Let us consider $w_i(0)$ to be a zero-mean white noise along the iteration axis i and a standard deviation $\sigma < \infty$. The sum on the right-hand side of (2.19)

$$
\sum_{j=0}^{i} \alpha^{i-j} w_j(0)
$$

can be viewed as the output of a first-order linear system

$$
\frac{1}{z - \alpha}
$$

driven by a random input $w_i(0)$. The power spectrum density of the sum is

$$
\frac{\sigma^2}{(z - \alpha)^2}
$$

which has a finite mean square value $\sigma^2/(1-\alpha)^2$. Therefore, the power spectrum density of $u_i(0)$ is proportional to σ, and bounded owing to the introduction of the forgetting factor α.

In real-time applications to motion systems, the zero initial condition can be easily guaranteed for higher-order states such as velocity, acceleration, as far as motion systems come to a full stop after every iteration. Thus, only the position state needs a perfect re-set, which can be easily achieved by means of a PI controller with a sufficiently long settling time.

2.3 Implementation Issues

2.3.1 Repetitive Control Tasks

ILC is essentially for tracking in a finite time interval, which is different from repetitive tasks – either tracking a periodic trajectory or rejecting a periodic exogenous disturbance. Repetitive control tasks are continuous in the time domain, such as the speed control of an electrical drive, or voltage control of a power supply. The common feature of this class of control problems is the presence of some undesirable cyclic behavior when the system works around an operating point, for instance a power supply generates a DC voltage, but perturbed by 50 Hz harmonics. Repetitive control was developed to deal with this class of control tasks, but encounters difficulty when a non-linear process is concerned. In fact, we have shown experimentally that a simple repetitive control law below can effectively deal with non-linear dynamics

$$u(t) = u(t-T) + \beta e(t) \qquad t \in [0,\infty), \tag{2.23}$$

where T is the period of the cyclic reference trajectory or cyclic disturbance. Unfortunately, we lack an effective analysis tool for repetitive control, as most repetitive control problems can only be analyzed in the frequency domain, despite the fact that the controller does perform well for non-linear processes.

Recently, three learning control methods were proposed for non-linear systems that perform repetitive control tasks. The first method uses the alignment condition (e) and employs Lyapunov methods in the analysis. It requires the plant to be parameterized, and all states to be accessible. This method is named as adaptive ILC [35, 37, 131, 152]. The second method extends (2.23) to the non-linear repetitive learning control laws in which periodic learning and robust control are integrated [30, 160]. The convergence analysis is conducted using the Lyapunov–Krasovskii functional. The third method is also a non-linear learning control method that stabilizes the process by a non-linear feedback controller, and meanwhile introduces a periodic signal [64]. Then, from [162] it is known that the closed loop will enter a cyclic steady state after a sufficiently long interval. Next, by changing the periodic signal, the system will enter another cyclic steady state after a sufficiently long interval. The control system will converge asymptotically if the periodic signal is updated, once after the long interval, according to the learning law (2.23).

According to the third method, the repetitive control process can be converted into the repeated control process, subsequetly the ILC analysis can be extended to repetitive control problems for a large class of non-linear processes. Note that the process is stable under feedback control but the undesirable cyclic behavior cannot be eliminated. Any change of the control signal will yield a new cyclic steady state. Thus, we can segment the horizon $[0, \infty)$ into a sequence of finite intervals $[iT, (i+1)T)$, $i = 0, 1, \cdots$. Let $\tau \in [0, T)$. Then, for any instant $t \in [0, \infty)$, there exists an interval $[iT, (i+1)T)$, such that $t = \tau + iT$. Denote $u_i(\tau) = u(\tau + iT)$, then $u(t - T) = u_{i-1}(\tau)$. By segmenting $\Delta y(t)$ in the same way as $u(t)$, the repetitive control law (2.23) can now be re-written as

$$u_{i+1}(\tau) = u_i(\tau) + \beta e_{i+1}(\tau) \qquad \tau \in [0, T). \tag{2.24}$$

Comparing with the current cycle ILC law (2.3), it can be seen that both share the same form. Next, T is sufficiently long so that the closed-loop response will enter the cyclic steady state at the end of each interval. Learning control can be updated between any two consecutive intervals. A remaining problem is how to choose the period T. In several learning applications, T was chosen to be the lowest common multiple among cyclic factors that should be rejected. A concern is that the system may not reach its cyclic steady state within a short T.

2.3.2 Robustness and Filter Design

In practical applications, the textbook-like ILC algorithms, such as (2.1) and (2.3), may not function properly. Often, we can observe such a V-phenomenon (drop-rise phenomenon) : the tracking error decreases at the first few iterations, but increases when the iterative learning process continues. A common explanation for such a V-phenomenon is the presence of non-repeatable factors in the learning process. Indeed, none of the practical processes are perfectly repeatable. The system response always contains both the repeatable and non-repeatable signals. Learning of non-repeatable signals could be dangerous, because ILC is essentially a point-wise integration along the iteration axis. However, this explanation is not conclusive. Let us see what kind of non-repeatable disturbances will result in an increasing learning error sequence. Consider the process $P(s)$ subject to an exogenous disturbance or measurement noise $D_i(s)$

$$Y_i = PU_i + D_i.$$

When the ILC law (2.1) is used, the ILC tracking error is

$$\begin{aligned} E_{i+1} &= Y_r - Y_{i+1} \\ &= Y_r - PU_{i+1} - D_{i+1} \end{aligned}$$

$$= Y_r - P(U_i + C_l E_i) - D_{i+1}$$
$$= (1 - PC_l)E_i - (D_{i+1} - D_i). \tag{2.25}$$

From (2.25), the repeatable perturbation $D_{i+1} = D_i$ can becompletely rejected. It was shown in the PCL discussion that, in the absence of the perturbations $D_{i+1} - D_i$, the convergence is determined by the factor

$$\|1 - PC_l\| \leq \gamma < 1$$

for frequencies of interest $\omega \in \Omega$, $\Omega = [\omega_a, \omega_b]$. Note that ILC is unable to reject non-repeatable perturbations. This is also true for any ILC method. Our concern is, in what circumstances will ILC lead to a divergent control sequence when non-repeatable D_i is present.

In (2.25), denote $\Delta_i = D_{i+1} - D_i$, we have

$$E_{i+1} = (1 - PC_l)^{i+1} E_0 + \sum_{j=0}^{i} (1 - PC_l)^{i-j} \Delta_j. \tag{2.26}$$

The V learning phenomenon can be partially explained from the relationship (2.26). The tracking error reduction at the first few iterations is owing to the attenuation of the initial error $(1 - PC_l)^{i+1} E_0$, whereas the subsequent increment of the error is due to the accumulation of $(1 - PC_l)^{i-j} \Delta_j$. Note that the summation term in (2.26) may remain bounded when the magnitude of $1 - PC_l$ is strictly below 1 for all frequencies. However, if the relative degree of PC_l is above 0, the magnitude of $1 - PC_l$ approaches 1 when the frequency ω is sufficiently high, leading to the occurrence of large tracking errors at high frequencies. The accumulation of Δ_i is because of the integral action in the ILC laws (2.1) and (2.3).

To prevent such an undesirable "integration" for non-repeatable components, filtering technology has been widely used in practice. Two commonly adopted modifications for ILC robustification are the forgetting factor and the low-pass filtering. Consider the PCL scheme with the forgetting factor $\alpha \in (0, 1]$

$$U_{i+1} = \alpha U_i + C_l E_i. \tag{2.27}$$

The tracking error is

$$E_{i+1} = Y_r - Y_{i+1}$$
$$= Y_r - PU_{i+1} - D_{i+1}$$
$$= Y_r - P(\alpha U_i + C_l E_i) - D_{i+1}$$
$$= (\alpha - PC_l)E_i - [D_{i+1} - \alpha D_i - (1 - \alpha)Y_r]. \tag{2.28}$$

Now, the convergence is determined by the factor $\alpha - PC_l$. When the relative degree of PC_l is above 0, the magnitude of $\alpha - PC_l$ approaches $\alpha < 1$. Thus, the forgetting factor can effectively prevent the high-frequency perturbation from accumulating. The main drawback of the forgetting factor is the non-discriminant attenuation to all frequencies including useful frequencies. To avoid the drawback of the forgetting factor, the constant forgetting factor α can be replaced by a frequency-dependent low-pass filter (LPF) $F(s)$

$$U_{i+1} = FU_i + C_l E_i, \tag{2.29}$$

where F has a cutoff frequency ω_c.

At the low-frequency band $\omega < \omega_c$, F is approximately equal to 1. The LPF modification works in the circumstances where D_i is negligible at the low-frequency band, and any attenuation of the reference trajectory Y_r should be avoided. At the high-frequency band $\omega > \omega_c$, the magnitude of F is less than 1, thus F works as a forgetting factor. In principle, if the power spectra of Y_r and D_i can be separated by ω_c, the LPF modification yields satisfactory responses. If the power spectra of Y_r and D_i overlap for frequencies $\omega > \omega_c$, perfect tracking of Y_r is not achievable and the introduction of LPF prevents the worst case of divergence. If the perturbation D_i contains low-frequency or DC components, it is not advisable to further lower the cutoff ω_c. Instead, the LPF should be revised such that the magnitude of F at the low-frequency band is strictly less than 1. If there is no prior knowledge about the power spectrum of D_i, the magnitude of the LPF should be chosen uniformly below 1, especially $|F(0)| < 1$. As a rule of thumb, $|F(0)| \in [0.9, 0.99]$ is recommended.

Finally, we discuss the structure of the LPF. Among numerous filters, Butterworth LPF is most widely used owing to the smooth frequency response, as shown in Fig. 2.7, and the adjustable stiffness in the transition band by the order of the filter. The frequency response function of the Butterworth LPF has the following form

$$|F(j\omega)|^2 = \frac{1}{1 + (j\omega/j\omega_c)^{2N}}, \tag{2.30}$$

where N is the order of the filter. From Fig. 2.7, it can been seen that Butterworth LPF has the pass band at $\omega = 0$ and the stop band at $\omega \to \infty$. At the cutoff frequency, $|F(j\omega_c)| = 1/\sqrt{2}$, regardless of the order of the filter. As the filter order increases, the transition from the pass band to the stop band gets steeper, namely approaching the ideal cutoff. In practical applications, the second-order Butterworth filter is often adequate to attenuate the high frequencies. Choosing the order $N > 2$ though can further improve the stop-band property, the phase lag will also increase and finally lead to a poor learning result.

2.3.3 Sampled-data ILC

ILC are implemented using microprocessors, not only because of the recent advances in microprocessor technology, but also because of the nature of ILC – use of a memory array to record the past control information. The sampled-data ILC shows unique characteristics, and attracted a number of researchers to explore and develop suitable ILC algorithms. Here,e briefly investigate three effects associated with the sampling.

The first and best known effect of sampling is the one-step sampling delay. In contrast to other control problems, ILC requires the identical initialization condition when aiming at the perfect tracking. Fortunately, PCL uses the past tracking error signals, hence the one-step delay can be easily compensated and the ILC law in the time domain is

$$u_{i+1}(k) = u_i(k) + \beta e_i(k+1), \tag{2.31}$$

where k is the sampling instance, β is a constant learning gain. Clearly, this one-step advance compensation cannot be implemented in CCL algorithms. Therefore in most ILC applications, a feedback C is first incorporated and the closed-loop transfer function is

$$G = \frac{PC}{1+PC},$$

where P and C are transfer functions in the z-domain. Next, the ILC is designed for the closed-loop system, accordingly, the learning convergence condition is

$$\|1 - GC_l\| \le \gamma < 1.$$

In a sense, this design is a PCCL, where the feedback C guarantees the closed-loop stability along the time axis, and the compensator C_l ensures the learning convergence along the iteration axis.

The second effect of sampling is related to the size of the sampling interval T_s, which directly determines the sampling frequency. From the sampling theorem, a smaller sampling interval is preferred and the sampling interval is a major factor that decides the tracking accuracy. However, a smaller sampling interval may degrade

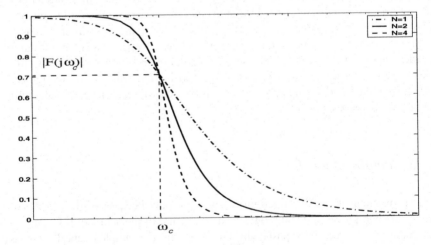

Fig. 2.7 Frequency response of Butterworth low-pass filters

the transient performance of ILC in the iteration domain. Consider a continuous-time process

$$\dot{\mathbf{x}} = A\mathbf{x} + Bu.$$

After sampling, the process is

$$\mathbf{x}(k+1) = \Phi\mathbf{x}(k) + \Gamma u(k),$$

$$\Phi = e^{A\Delta} \qquad \Gamma = \int_0^\Delta e^{A\tau} d\tau.$$

When A is asymptotically stable, all the eigenvalues of Φ are within the unit circle. The larger the T_s, the closer are eigenvalues of Φ to the origin. Recent research [1] shows that a monotonic convergence along the iteration axis can be achieved if $\|\Phi\|$ is sufficiently small. Therefore, a smaller sampling interval can improve the tracking accuracy in the time domain, whereas a larger sampling interval could improve the transient response of ILC in the iteration domain. A multi-rate ILC controller can be considered. For the closed-loop feedback the minimum available sampling interval should be used. For the learning loop we can first use a larger sampling interval to avoid a poor transient response, and reduce the sampling interval when the tracking error does not decrease further as the iteration proceeds.

The third effect of the sampling is associated with the relative degree of the original continuous-time process. It is known that sampling with zero-order hold can reduce the relative degree from $n > 1$ in continuous time to 1 in the sampled data. As a consequence, it is adequate to use the ILC law (2.31) even if the original continuous process has a relative degree above 1. This property is highly desirable, as only one-step advance compensation is required. In comparison, if the ILC is designed in continuous time, nth-order differentiation is necessary, but the implementation of such a high-order differentiator is not an easy task, especially when measurement noise is present. Another effect, a side effect, from using a smaller sampling interval is the generation of a non-minimum phase model. It has been shown in [10], that a non-minimum phase discrete-time process will be generated if the original continuous-time process has a relative degree above 2 and the sampling interval is sufficiently small. As a result, perfect tracking is not possible. In such circumstances, multi-rate CCL is an appropriate choice, and the sampling interval for the learning updating mechanism should be chosen as long as possible as long as the tracking accuracy is satisfied.

2.4 Conclusion

This chapter focused on several important issues linked to the basic ILC configurations, GLC conditions, identical initialization conditions, filtering and sampling. There are many issues involved in the ILC implementation and they could not be covered in this chapter. Some of these issues will be further addressed in subsequent chapters when the real-time implementation problems are explored. Some other issues were explored in recent ILC literature, and are surveyed in the preceding chapter. No doubt many are still open problems and wait for us to solve in the future.

Chapter 3
Robust Optimal ILC Design for Precision Servo: Application to an XY Table

Abstract This chapter addresses the high-precision servo-control problem using the iterative learning control techniques. Instead of seeking an accurate model of servomechanism, servo tracking performance is enhanced via repeated learning. To facilitate the sampled learning control system design, we propose two optimal objective functions. The first objective function is to maximize the frequency range in which learning converges, and subsequetly enhance the system robustness. The second objective function is to search the fastest learning convergence speed in the iteration domain. Under a unified optimal design, we then compare three representative iterative learning control algorithms PCL, CCL and PCCL, exploit their suitability for the servo implementation. We further elaborate on the issue of robust optimal design, which seeks the fastest learning convergence under the worst-case parametric uncertainties. The experimental results show that the high-precision tracking performance is achieved via iterative learning, despite the existence of miscellaneous non-linear and uncertain factors.

3.1 Introduction

Precision servo has been widely applied in precision manufacturing industry, such as wafer process, IC welding process, polishing process of airplane parts, the production of miscellaneous precision machine tools, *etc*. The key feature of these processes can be characterized by 1) the high-precision specification, 2) the repeated task, and 3) finite time interval for each operation cycle.

In many existing servo-control methods, dominated by feedback techniques and characterized by the asymptotic convergence, it will be hard to achieve perfect tracking within a rather short operation cycle. Moreover, the same tracking performance will be observed no matter how many times the servo-task repeats. On the other hand, the repeated action of a dynamic process would provide the extra information from past control input and tracking error profiles. If a servomechanism is capable of actively using the extra information from the past control operations, a better tracking performance can be expected.

The ILC technique, which employs memory components to store and use the past control signals to update the control action of the present operation cycle, provides a new paradigm to solve the high precision-servo problems under a repeatable control environment, [4, 13, 14, 19, 22, 28, 37, 59, 67, 76, 84, 89, 95, 100, 107, 144, 148]. When a model-based servo-control is designed to pursue a high-precision tracking performance, the modelling accuracy becomes the bottleneck. Instead of trying to model various imperfect factors in a servo -ystem, which is almost impossible due to the existence of non-linearities and uncertainties, an ILC servomechanism directly "learns" the necessary control action via the repeated process. This is considered as one of the typical advantages of intelligent control methods: requiring less *a priori* knowledge on the process model, and achieving a progressive improvement through a well-designed learning mechanism.

In this work an XY table driven by two DC servo-motors is considered. A DC motor is in general modelled by a second -rder transfer function, and this model can work reasonably well if the control task requires a moderate tracking precision. In real life there are many imperfect factors either ignored or extremely hard to model: the deadzone, the viscous and static frictions in mechanical contacts such as the ball screw, the light coupling between two axes, the eccentricity in bearings, the elasticity in torque transmission devices, torque ripples from the power amplifier, neglected dynamics such as sensor dynamics, electrical drive dynamics, and motor electrical dynamics, *etc.* These imperfect factors, though tiny, will hinder any attempt to further improve the servo-precision, and become predominant when a high-precision tracking is required. In such circumstances, ILC provides a unique alternative: capturing the inherently repeated or the operation-wise invariant components in the control signals, and revise the control input accordingly. Non-repeatable components, on the other hand, can be easily rejected by incorporating appropriate band pass-filters.

By virtue of the learnability of ILC, the second-order DC motor model, used as the nominal part, will be adequate for ILC servo design. The imperfect modelling factors, which do not affect the system convergence but the perfection of tracking, will be "learned" and "rejected" through iterative learning. In the ILC literature, the main attention has been on the convergence analysis in the iteration domain. Our first objective in this work is to design an optimal ILC servo based on the nominal model. The first step is to model the servo, which is done by means of least squares estimation in the frequency domain. The second step is to choose appropriate performance indices. Consider the importance of the robustness in the servomechanism, the first index is the maximization of the frequency range such that the ILC servomechanism converges within the range. Owing to the nature of digital control, the maximum frequency is one half of the Nyquist frequency. Another performance index, from the practical point of view, is chosen to be the learning convergence speed in the iteration domain. Based on the optimal criteria, the third step is to search the optimal control parameters. Since both criteria are highly non-linear in the parametric space, it is not possible to solve the optimization problem in the closed form. In such circumstances numerical methods and evolutionary computation methods can be used.

Our second objective in this work is to present, analyze and apply the three most fundamental and representative ILC algorithms to the high-precision control of a DC servomechanism. Since the servo nominal model is simply second order, it is natural to choose a PD-type control scheme accordingly. Without any further detailed model knowledge, it would be difficult to design a complex or non-linear controller. Again as we indicated, it is extremely difficult to model various imperfect factors and it would be a much easier job if we can let the servomechanism automatically "learn" to compensate or reject the effect of those factors when the task repeats, which is the ultimate goal of integrating the servomechanism with ILC in this work. In order to conduct a fair comparison, all three learning algorithms are equally optimized in the same parametric space, based on the proposed two optimal criteria.

A sampled-data learning mechanism updated with the previous cycle tracking error (PCL) is first studied. This ILC algorithm, like the majority of ILC algorithms, works in an open-loop fashion in the time domain. The computed optimal design shows that PCL has a very slow convergence speed. This, together with the open-loop nature, indicates that PCL is not suitable for high-precision servo-control. In fact, PCL presents an experimentally divergent behavior, due to its sensitivity to the imperfect factors.

To overcome the shortcoming of PCL, a sampled-data learning mechanism updated with the current cycle tracking error (CCL) is considered. The closed-loop nature of CCL enhances the robustness of the servomechanism against imperfect factors. Using the same optimal design, CCL shows a much faster convergence speed than that of PCL. The experimental result verifies that, in spite of the imperfect factors, CCL can effectively reduce the tracking error by 98% within 10 iterations.

The integration of both PCL and CCL (PCCL) provides more degrees of freedom in the parametric space and thus it offers the possibility of achieving a better tracking performance. The computed optimal design shows that PCCL can further improve the convergence speed by 10%–20%, and the experimental result verifies that PCCL can effectively reduce the tracking error by more than 98% within 7 iterations.

Finally, we consider the robust optimal ILC design problem where the servo nominal model itself may undergo certain parametric variations. Such parametric variations are observed in the experiments when the amplitude scales of servo input signals vary with respect to different target trajectories. This implies there may exist an unmodelled non-affine factor in the servomechanism input. The robust optimal design method is employed in the PCCL algorithm by taking the servo-parametric variations into account, and its effectiveness is verified via the experimental results.

The chapter is organized as follows. Section 3.2 describes the model of the DC servomechanism that drives a XY table, and the design consideration of the optimal ILC. Sections 3.3, 3.4, and 3.5 detail the optimal designs of the PCL, CCL and PCCL algorithms respectively, and present experimental results for comparison. In Sect. 3.6, the robust optimal design for PCCL is addressed in the presence of the parameters variations. Finally, Sect. 3.7 concludes the chapter.

3.2 Modelling and Optimal Indices

In this section the control process and the control task will be detailed.

3.2.1 Experimental Setup and Modelling

The hardware layout of the XY table is shown in Fig 3.1. In each axis, there is a brush -ype SEM's MT22G2-10 DC servo-motor driving the XY table through a ball screw. The motor has the maximum speed of 5000 rpm and the peak stall torque of 4 N m. An optical sensor attached to the motor measures the position with 4000 pulses per revolution. A pulse-width-modulation (PWM) power amplifier, which could produce up to 7 A continuous current, is used in current mode to drive the DC motor. The controller used for algorithm execution is designed in a Matlab/Simulink environment while the complied program is sent to a TMS 320 DSP chip for data acquisition and real-time control. The sampling period is set to be 2 ms.

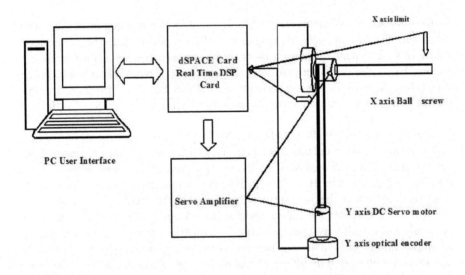

Fig. 3.1 Block diagram of the system

By ignoring the motor electrical dynamics, the relation between the linear motion of the XY table along each axis and the motor input current can be approximated as a second-order nominal model

$$P(s) = \frac{a}{bs^2 + s}. \tag{3.1}$$

The two lumped parameters a and b in general depend on various system parameters such as the motor torque coefficient, the load and motor rotor inertia, viscous coefficient, ball screw pitch, and the gear ratio of the belt coupling. Note that we ignored the electrical dynamics under the assumption that the current mode can perform well. These system parameters are either difficult to calibrate or deviate from the rated values. Thus, the first step in the control design is to estimate the two parameters a and b in the frequency domain with the help of least squares estimation and FFT techniques. After extensive experiments, the estimated parameters are $a = 2.844$ and $b = 0.354$ for the X-axis, and $a = 2.1$ and $b = 0.31$ for the Y-axis. For simplicity in the remaining part of the chapter we use $P(z)$ to denote the z-domain transfer functions of either the X-axis or Y-axis.

3.2.2 Objective Functions for Sampled-data ILC Servomechanism

Consider a typical control task for an XY table: drawing a circle in a specified time interval $[0, 5]$, with the motion pattern along each axis described by

$$x_r = 0.05\sin(0.0121t^5 - 0.1508t^4 + 0.5027t^3) \text{ m}$$
$$y_r = 0.05\cos(0.0121t^5 - 0.1508t^4 + 0.5027t^3) \text{ m}, \qquad (3.2)$$

and shown in Fig. 3.2. The purpose of incorporating the 5th-order polynomial is to let the trajectory start with a zero speed and zero acceleration, and subsequently avoid the need for an impulse signal.

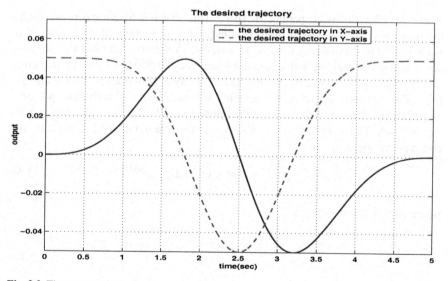

Fig. 3.2 The target trajectory

In terms of the nominal model and its discretization $P(z)$, it is appropriate to choose a PD-type digital controller

$$C(z) = k_p + k_I \frac{z-1}{T_s z}, \tag{3.3}$$

where k_p, k_I and $T_s = 2$ ms are proportional gain, derivative gain and sampling period respectively. In order to guarantee a smooth and unsaturated control, and take into consideration the hardware limits especially the upper and lower current limits $[-0.5, 0.5]$ of the servo motor, the ranges of PD gains are confined to $K \in [0, 50]$ and $T \in [0, 5]$ in this work. Experiments show that the tracking error is about 3×10^{-3} m. Here, our control objective is to further reduce the tracking error to the scale of 10^{-4} or even less, say 10^{-5} m, through learning.

Let the control process repeat over a finite time interval $[0, 5]$ seconds. A typical ILC can be written as

$$U_{i+1}(z) = U_i(z) + f(E_{i+1}(z), E_i(z), \mathbf{p}), \tag{3.4}$$

where the subscript i denotes the iteration number, $f(\cdot, \cdot, \cdot)$ is a smooth function of the arguments, $\mathbf{p}$ denotes all the controller parameters, $E_i(z)$ is the Z-transform of the tracking error $e_i(k)$ of either the X-axis $x_r(k) - x_i(k)$ or the Y-axis $y_r(k) - y_i(k)$, and k is the sampled instances.

The optimal sampled ILC design can be formulated first as

$$J = \begin{cases} \displaystyle\max_{\mathbf{p} \in \mathscr{P}} \; \omega_o \\ s.t. \; \dfrac{|E_{i+1}(z)|}{|E_i(z)|} = \rho(z) < 1 \quad \forall\, \omega \leq \omega_o \text{ and } z = e^{j\omega T_s}. \end{cases} \tag{3.5}$$

where $\mathscr{P}$ is the admissible parametric space. This objective function is to maximize the frequency range in which the ILC convergence is guaranteed. In a sense, this is equivalent to maximizing the ILC robustness. Considering the fact that the servomechanism is working with a sampling frequency of 500 Hz, we need only consider frequencies ω_o upto 250 Hz, one half of the Nyquist frequency. This leads to a possibility that there may exist a non-empty subspace $\mathscr{P}_s \subset \mathscr{P}$ such that $\forall \mathbf{p} \in \mathscr{P}_s$, $\omega_o = 250$ Hz. We can further exploit the optimality of ILC in the reduced parametric space $\mathscr{P}_s$. The second objective function considered in this work is the learning convergence speed

$$J = \min_{\mathbf{p} \in \mathscr{P}_s} \rho(z) \quad \forall\, \omega \leq \omega_o \text{ and } z = e^{j\omega T_s}, \tag{3.6}$$

where $\rho(z)$ is defined in (3.5).

3.3 Optimal PCL Design

The block diagram of PCL is shown in Fig. 2.1. PCL works on the basis of the previous cycle information, including both previous cycle error and controller input stored in the memory. All information in the ith iteration could be utilized in the control of $(i+1)$th iteration. A simple sampled-data ILC with the PCL algorithm is constructed as

$$U_{i+1}(z) = U_i(z) + C_l(z)zE_i(z). \tag{3.7}$$

where C_l is an appropriate compensator or filter that is causal. The extra z operating on $E_i(z)$ denotes a left shift of the error signal, or a one-step-ahead shift. The purpose is to compensate the delay incurred by the sampling mechanism. Consider the Y-axis (all results hold equally for the X-axis)

$$E_i(z) = Y_r(z) - P(z)U_i(z), \tag{3.8}$$

the tracking error at the $(i+1)$th iteration can be represented as

$$\begin{aligned} E_{i+1}(z) &= Y_r(z) - Y_{i+1}(z) \\ &= Y_r(z) - P(z)[U_i(z) + C_l(z)zE_i(z)] \\ &= [1 - zP(z)C_l(z)]E_i(z), \end{aligned} \tag{3.9}$$

which leads to the convergence condition of PCL

$$|1 - zC_l(z)P(z)| = \rho(z) < 1. \tag{3.10}$$

Based on the second-order model $P(z)$, we choose the compensator $C_l(z)$ in the form of a discretized PD controller,

$$C_l(z) = k_p + k_d \frac{z-1}{T_s z}, \tag{3.11}$$

where k_p is the proportional gain and k_d is the derivative gain. Note that most ILC algorithms require a relative degree of 0, whereas the plant model (3.1) has a relative degree of 2. It is desirable to have a second-order derivative signal, namely the acceleration information. However, the servomechanism only provides the position sensor. By twice numerical difference the measurement noise will be amplified by $T_s^{-2} = 2.5 \times 10^5$! Thus, as a tradeoff, a PD-type compensator is appropriate in such circumstances.

The parametric space is $\mathscr{P} = [0, 50] \times [0, 5]$. Substituting $C_l(z)$ into (3.10) yields the PCL convergence condition

$$\rho(e^{j\omega T_s}) = \left| 1 - e^{j\omega T_s} \left(k_p + k_d \frac{e^{j\omega T_s} - 1}{T_s e^{j\omega T_s}} \right) P(e^{j\omega T_s}) \right| < 1. \tag{3.12}$$

The optimal design according to the first objective function (3.5) can be written as

$$J_1 = \begin{cases} \max\limits_{(k_p,k_d)\in\mathscr{P}} \omega_o \\ s.t.\ \rho(e^{j\omega T_s}) < 1 \quad \forall\ \omega \le \omega_o. \end{cases} \tag{3.13}$$

Through computation, it is found that there exists a non-empty subset $\mathscr{P}_s \in \mathscr{P}$ such that the first objective function (3.13) can reach its maximum $\omega_o = 250$ Hz. Hence the second objective function can be optimized with feasible solutions

$$J_{PCL} = \min\limits_{(k_p,k_d)\in\mathscr{P}_s} \left| 1 - e^{j\omega T_s} \left(k_p + k_d \frac{e^{j\omega T_s} - 1}{T_s e^{j\omega T_s}} \right) P(e^{j\omega T_s}) \right| \tag{3.14}$$
$$\forall\ \omega \le 250\ \text{Hz}.$$

By solving (3.14) numerically for both the X-axis and Y-axis, the optimal values are

$$\text{X} - \text{axis}: \begin{cases} k_p = 0.54 \\ k_d = 0.54 \\ J_{PCL} = 0.9919 \end{cases} \qquad \text{Y} - \text{axis}: \begin{cases} k_p = 0.32 \\ k_d = 0.32 \\ J_{PCL} = 0.9959. \end{cases} \tag{3.15}$$

ILC design operates in an off-line fashion. Therefore, computation time in searching optimal controller parameters will not be our concern in this work. Once the criteria are given, we can always come up with a satisfactory solution either using numerical computation or evolutionary computation methods. In this work we use a genetic algorithm to search for the optimal solution.

The convergence conditions of PCL in the frequency domain are shown in Fig. 3.3. It can be seen that in both the X-axis and Y-axis the optimal convergence speeds J_{PCL}, which are defined as the ratio between any two consecutive iterations, are close to 1. This implies that the PCL convergence will be very slow. Moreover, the learning convergence condition may be easily violated, as any modelling uncertainties may alter the system transfer function at a certain frequency band.

Experimental results are demonstrated in Fig. 3.4, where the horizon is the iteration number and the vertical is the maximum tracking error defined as $e_{ss} = \max\limits_{t\in[0,5]} |e_i(t)|$.

This shows that the tracking error does not converge in the iteration domain, though the theoretical calculation based on the nominal model indicates a convergent behavior. In this work, throughout all experiments, a Butterworth filter with 20 Hz center frequency is added in order to avoid differentiating the high-frequency measurement noise.

3.4 Optimal CCL Design

The block diagram of CCL is shown in Fig. 2.2. Comparing with the PCL configuration in Fig. 2.1, the fundamental difference is the incorporation of the current cycle feedback loop. The first advantage we can expect is the enhancement of the system robustness property in the time domain. In the following, we will show that the CCL algorithm also speeds up learning in the iteration domain.

The CCL is constructed as follows

$$U_{i+1}(z) = U_i(z) + C(z)E_{i+1}(z) \tag{3.16}$$

where C is the CCL controller in the z-domain, and $E_{i+1}(z) = Y_r(z) - P(z)U_{i+1}(z)$. In terms of the second-order nominal model, a PD controller $C(z)$

$$C(z) = k_p + k_d \frac{z-1}{T_s z} \tag{3.17}$$

is chosen with k_p and k_d the proportional and derivative gains, respectively. The CCL convergence condition can be easily derived as follows

$$
\begin{aligned}
E_{i+1}(z) &= Y_r(z) - P(z)U_{i+1}(z) \\
&= Y_r(z) - G(z)U_i(z) - P(z)C(z)E_{i+1}(z) \\
\Rightarrow \quad & [1 + P(z)C(z)]E_{i+1}(z) = E_i(z).
\end{aligned}
\tag{3.18}
$$

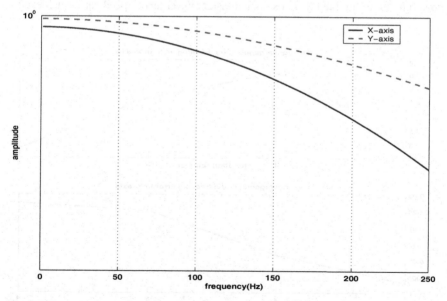

Fig. 3.3 The convergence conditions of PCL algorithm in the frequency domain

Thus, $e_i(k)$ converges in the iteration domain if

$$\rho(z) = \frac{|E_{i+1}(z)|}{|E_i(z)|} = \frac{1}{|1+P(z)C(z)|} < 1. \qquad (3.19)$$

From the frequency-domain viewpoint, CCL can track the desired trajectory precisely if

$$\frac{1}{|1+C(e^{j\omega T_s})P(e^{j\omega T_s})|} < 1 \quad \forall \, \omega \leq \omega_o, \qquad (3.20)$$

and if the desired trajectory has a bandwidth less than ω_o. Analogous to PCL analysis and design, it is adequate if we can design the PD gains to make the maximum $\omega_o = 250$ Hz. Consequently, the first optimal objective is

$$J_1 = \begin{cases} \displaystyle \max_{(k_c,k_d)\in\mathscr{P}} \omega_o \\[2mm] s.t. \; \dfrac{1}{\left|1+(k_p+k_d\dfrac{e^{j\omega T_s}-1}{T_s e^{j\omega T_s}})P(e^{j\omega T_s})\right|} < 1 \quad \forall \, \omega \leq \omega_o. \end{cases} \qquad (3.21)$$

The existence of an analytic solution in (3.21) is in general very difficult to verify. The search for feasible solutions of (3.21), thus, is again conducted numerically. From the computation, there exists a large subset $\mathscr{P}_s \subset \mathscr{P}$ such that $\omega_o = 250$ Hz. Analogous to the PCL design, we now search for the learning control gains that can optimize the convergence speed in the iteration domain. The objective is now to find the suitable (k_p, k_d) so that a faster convergence speed can be achieved,

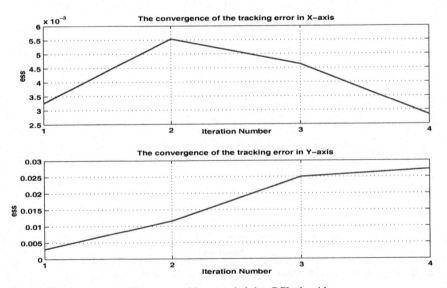

Fig. 3.4 The maximum tracking errors with a sampled-data PCL algorithm

while retaining the convergence within the frequency range of 250 Hz. The second objective function can be represented as

$$J_{CCL} = \min_{(k_p, k_d) \in \mathscr{P}_s} \frac{1}{\left| 1 + \left(k_p + k_d \dfrac{e^{j\omega T_s} - 1}{T_s e^{j\omega T_s}} \right) P(j\omega T_s) \right|} \quad \forall\, \omega \le 250\,\text{Hz.} \quad (3.22)$$

By solving (3.22) for both the X-axis and Y-axis, the optimal values are

$$\text{X} - \text{axis}: \begin{cases} k_p = 50.00 \\ k_d = 5.00 \\ J_{CCL} = 2.4120 \times 10^{-4} \end{cases} \qquad \text{Y} - \text{axis}: \begin{cases} k_p = 50.00 \\ k_d = 5.00 \\ J_{CCL} = 2.8138 \times 10^{-4}. \end{cases} \quad (3.23)$$

The convergence conditions of CCL in the frequency domain are shown in Fig. 3.5 and the convergence of the tracking errors in the X-axis and Y-axis is shown in Fig. 3.6.

It turns out that the best solution is when the PD gains take the maximum values in $\mathscr{P}$. Hence, a high-gain feedback also improves the learning performance in the iteration domain, as far as the precision servomechanism is concerned. Note that the tracking error at the first iteration is the result of PD control. The tracking error has been successfully reduced by about 98% to the scale of 5×10^{-5} after ten iterations. A question is, can we use higher-gains feedback alone to obtain the same level of tracking error? In fact, in the experiments we found that, by increasing the PD gains to fourteen times the K_c^* and T_c^*, that is, $k_p = 7000$ and $k_d = 700$, the tracking error can be reduced to about 2×10^{-4}. However, on further increasing control gains the tracking error goes up again because the servo becomes quite oscillatory due to the limited sampling frequency. Obviously it is not practical to use such high feedback gains. The servomechanism becomes very easy to get saturated and subsequetly

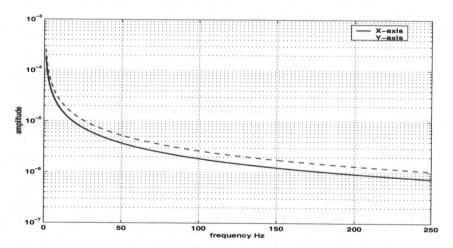

Fig. 3.5 The convergence conditions of the CCL algorithm in the frequency domain

lose the robustness due to the open-loop nature. One promising advantage of the ILC-based servomechanism, as clearly demonstrated here, is the ability to generate high-precision control with the low-gain feedback.

3.5 Optimal PCCL Design

The PCCL algorithm combines both sampled-data PCL and CCL, aiming at a better tracking performance. The block diagram of PCCL is shown in Fig.2.3

The learning control law can be written as

$$U_{i+1}(z) = U_i(z) + C(z)E_{i+1}(z) + C_l(z)zE_i(z) \tag{3.24}$$

which is the combination of two ILC laws (3.7) and (3.16).

The convergence condition of PCCL is derived as follows

$$
\begin{aligned}
E_{i+1}(z) &= Y_r(z) - P(z)U_{i+1}(z) \\
&= Y_r(z) - P(z)U_i(z) - P(z)C(z)E_{i+1}(z) - zP(z)C_l(z)E_i(z) \\
\Rightarrow \quad & [1 + P(z)C(z)]E_{i+1}(z) = [1 - zP(z)C_l(z)]E_i(z). \\
\Rightarrow \rho(z) &= \frac{|1 - zP(z)C_l(z)|}{|1 + P(z)C(z)|} < 1. \tag{3.25}
\end{aligned}
$$

It is obvious that both PCL and CCL are subsets of PCCL. If $C(z) = 0$, the PCCL algorithm leads to PCL, while if $C_l(z) = 0$, the PCCL algorithm leads to CCL. By

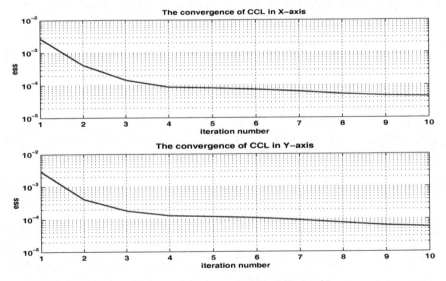

Fig. 3.6 The maximum tracking errors with a sampled-data CCL algorithm

integration, the PCCL (3.24) provides two degrees of freedom in the controller, one from feedforward (PCL) and one from feedback (CCL). Hence, a better control performance can be expected.

As was discussed in preceding sections, PD-type controllers (3.11) and (3.17) are employed. The control parametric space is now $(k_p^\circ, k_d^\circ, k_p, k_d) \in \mathscr{P}^2$, where k_p° and k_d° are PCL gains and k_p and k_d are CCL gains. Since both PCL and CCL meet the frequency requirement $\omega_o = 250$ Hz in some reduced subspaces $\mathscr{P}_s$ of $\mathscr{P}^2$, we can straightforwardly search for gains that maximize the learning convergence speed

$$J_{PCCL} = \min_{(k_p^\circ, k_d^\circ, k_p, k_d) \in \mathscr{P}_s^2} \frac{\left| 1 - e^{j\omega T_s} \left(k_p^\circ + k_d^\circ \dfrac{e^{j\omega T_s} - 1}{T_s e^{j\omega T_s}} \right) P(e^{j\omega T_s}) \right|}{\left| 1 + \left(k_p + k_d \dfrac{e^{j\omega T_s} - 1}{T_s e^{j\omega T_s}} \right) P(e^{j\omega T_s}) \right|}$$

$$\forall \omega \leq 250 \text{ Hz.} \tag{3.26}$$

The optimal solution of (3.26) is

$$X : \begin{cases} k_p^\circ = 5.0500 \times 10^{-3}; \\ k_d^\circ = 3.917753 \\ k_p = 5.00 \\ k_d = 50.00 \\ J_{PCCL} = 2.2013 \times 10^{-4}, \end{cases} \qquad Y : \begin{cases} k_p^\circ = 3.3900 \times 10^{-3} \\ k_d^\circ = 0.6410; \\ k_p = 50.00 \\ k_d = 5.00; \\ J_{PCCL} = 2.4136 \times 10^{-4}. \end{cases} \tag{3.27}$$

Comparing with CCL (3.23), we can see that the computed convergence speed of PCCL is about 10% – 20% faster than that of CCL.

The convergence of PCCL in the frequency domain is shown in Fig. 3.7. The tracking performance in the iteration domain is shown in Fig. 3.8, which exhibits a reduction of the tracking error from 3×10^{-3} to 5×10^{-5} after seven iterations. In other words, PCCL can successfully reduce the tracking error by 98%. Comparing with CCL, PCCL reaches the same scale of precision with fewer iterations, about 20% less, which is consistent with the ratio between computed results of J_{CCL} and J_{PCCL}. Figure 3.9 shows the control input signals of the X-axis at the seventh iteration. The big "jumps" correspond to the deadzone, that is, the ILC algorithm tries to automatically compensate the unknown deadzone as much as possible. We will further explore this property in the next chapter. It is also clear that the existence of high-frequency components is inevitable. Since a learning servomechanism may accumulate these high-frequency components, in practice learning should be terminated after several iterations. This can be implemented by simply adding a rule to check whether the tracking specification has been reached, and perhaps adding one more rule to cease learning updating compulsorily after a certain number of iterations if the tracking error still cannot meet the requirement. It is noted that the actual learning convergence speeds of all three ILC algorithms deviate a great deal from the computed ones. This shows that the impact from imperfect factors could be strong in the scenario of high-precision control, and a nominal model can hardly capture those factors. Nevertheless learning control is able to compensate those imperfect factors to a satisfactory level.

3.6 Robust Optimal PCCL Design

In the experiment, it is found that even the nominal model may vary when the servo input signals have different amplitude scales. This indicates the existence of a non-affine-in-input factor in the servomechanism. Since only a second-order model (3.1) is considered, the variations are reflected in the two plant parameters a and b. By feeding sinusoidal signals with different amplitudes to the servomechanism, the frequency-domain least squares estimation gives corresponding values of a and b as shown in Table 3.1.

Table 3.1 Experimental results of model variations

Amplitude of servo input	X-axis		Y-axis	
	a	b	a	b
0.1	1.066	0.1876	1.1746	0.2687
0.15	1.6561	0.2625	2.0523	0.3138
0.2	2.844	0.354	2.1	0.31

Obviously in the previous optimal designs and comparisons only the third case is considered. It is necessary to extend the optimal design to a robust optimal design by taking the nominal model variations into account. We have shown that PCCL is, in general, superior to PCL and CCL, thus here we focus on the robust optimal design

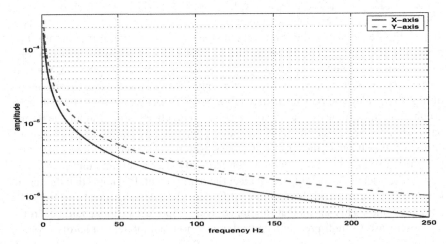

Fig. 3.7 The convergence conditions of a sampled-data PCCL algorithm in the frequency domain

of PCCL. Let $\mathscr{D}$ denote a parametric space of either the X-axis $[1.066, 2.844] \times [0.1876, 0.364]$ or the Y-axis $[1.1746, 2.1] \times [0.2867, 0.3138]$. The robust optimal design is implemented by revising the two objective functions. The first objective function (3.5), originally a max operation, now should be a max-min operation

$$J_1 = \begin{cases} \max\limits_{(k_p^\circ, k_d^\circ, k_p, k_d) \in \mathscr{P}^2} \min\limits_{(a,b) \in \mathscr{D}} \omega_o \\ s.t. \; \rho(z) < 1 \quad \forall \; \omega \leq \omega_o \; \text{and} \; z = e^{j\omega T_s}, \end{cases} \tag{3.28}$$

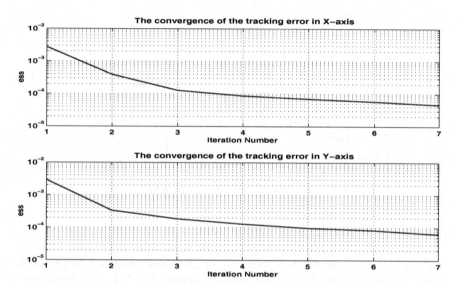

Fig. 3.8 The maximum tracking errors with a sampled-data PCCL algorithm

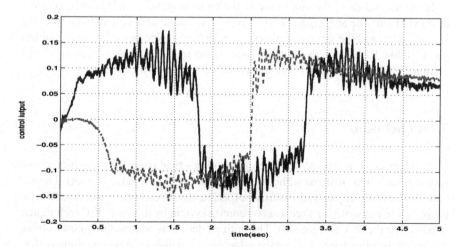

Fig. 3.9 The control profiles in the X-axis and Y-axis of a PCCL algorithm at the seventh iteration

where $\rho(z)$ is defined in (3.25). Likewise, if the objective function (3.28) has feasible solutions $\omega_o = 250$ Hz in a reduced control parametric space $\mathscr{P}_s^2 \subset \mathscr{P}^2$, we can proceed to the optimal convergence speed, and the corresponding objective function (3.6), originally a min operation, now should be a min-max operation

$$J_{PCCL} = \min_{(k_p^\circ, k_d^\circ, k_p, k_d) \in \mathscr{P}_s^2} \max_{(a,b) \in \mathscr{D}} \frac{\left| 1 - e^{j\omega T_s} \left(k_p^\circ + k_d^\circ \frac{e^{j\omega T_s} - 1}{T_s e^{j\omega T_s}} \right) P(e^{j\omega T_s}) \right|}{\left| 1 + \left(k_p + k_d \frac{e^{j\omega T_s} - 1}{T_s e^{j\omega T_s}} \right) P(e^{j\omega T_s}) \right|} < 1$$

$$\forall \omega \leq 250 \text{ Hz.} \tag{3.29}$$

Here, the servo model $P(z)$ is a function of parameters a and b, so is the quantity $\rho(z)$. Clearly, by adding extra min or max operations to the objective functions, the worst case in the parametric space $\mathscr{D}$ is taken into consideration. This warrants a robust optimal design with respect to unknown parametric variations.

The numerical solution of the robust optimal design of PCCL is

$$X - \text{axis}: \begin{cases} k_p^\circ = 0.0965; \\ k_d^\circ = 0.0855 \\ k_p = 50.00 \\ k_d = 5.00 \\ J_{PCCL} = 6.3512 \times 10^{-4}, \end{cases} \qquad Y - \text{axis}: \begin{cases} k_p^\circ = 0.012; \\ k_d^\circ = 1.9789 \\ k_p = 50.00; \\ k_d = 5.00; \\ J_{PCCL} = 3.0791 \times 10^{-3}. \end{cases} \tag{3.30}$$

Here, J_{PCCL} is the worst-case convergence speed.

Let us compare two cases. In the first case the target trajectory (3.2) is used, the resulting servo input amplitude is about 0.2, and the parameters a and b correspond to the last row of Table 3.1. In such case the optimal PCCL design should give the best result. Nevertheless, experimental results in Fig. 3.10 show that the robust optimal PCCL can work equally well and achieve almost the same performance.

In the second case, the amplitude of the target trajectory (3.2) is reduced from 0.05 to 0.03 m. The amplitude of servo-control signals is also scaled down from 0.2 to about 0.15, and the servo parameters a and b likely correspond to the second last row of Table 3.1. Again, the experimental results in Fig. 3.11 show that the robust optimal PCCL works equally well or better than the optimal PCCL.

3.7 Conclusion

In this chapter we addressed one important and practical issue: can a high-precision servomechanism be realized without using an equally high-precision model. Through both theoretical analysis and intensive experimental investigation, we demonstrate the possibility of realizing such a servo-control system by means of ILC techniques, in particular the PCCL algorithm. The optimal design as well as the robust optimal design, on the other hand, warrant the achievement of either the best tracking perfor-

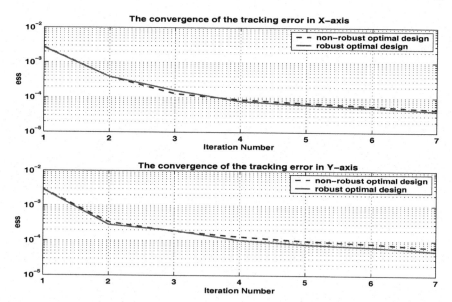

Fig. 3.10 Comparison of tracking errors for case 1

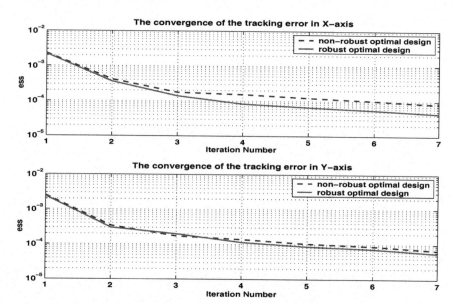

Fig. 3.11 Comparison of tracking errors for case 2

mance or the worst-case optimal performance. Finally, it is worth pointing out, that an ILC-based servomechanism possesses very useful characteristics: it uses only a low-gain feedback but achieves a high-precision tracking performance owing to the learning nature.

Chapter 4
ILC for Precision Servo with Input Non-linearities: Application to a Piezo Actuator

Abstract Most ILC schemes proposed hitherto were designed and analyzed without taking the input non-linearities, such as deadzone, saturation, backlash, into account. These input non-linearities are non-smooth and non-affine-in-input factor widely existing in actuators or mechatronics devices. They give rise to extra difficulty in control due to the presence of singularity or non-uniqueness in the system input channels. In this chapter, we focus on input deadzone, and show that ILC algorithms remain effective in general for systems with input non-smooth non-linearities that could be unknown and state dependent. In addition, in this chapter we will also briefly discuss the input saturation and input backlash.

4.1 Introduction

Input deadzone is a kind of non-smooth and non-affine-in-input factor widely existing in actuators or mechatronics devices because of friction. It gives rise to extra difficulty in control due to the presence of a singularity in the system input channels. Due to the wide presence in many practical processes, the input deadzone problem has been studied by many researchers. Some useful techniques for overcoming deadzone are variable structure control [139] and dithering [29]. Motivated by pursuing better control performance, several adaptive inverse approaches were proposed [105, 127, 128, 129], which employed an adaptive inverse to cancel the effect of an unknown deadzone. Recently, soft computing such as fuzzy logic and neural-network-based control algorithms have also been applied to handle problems relevant to deadzones. Fuzzy-logic-based controllers were developed in [63, 75]. Fuzzy pre-compensation schemes for PD controllers were proposed in [62]. Neural-network schemes [16, 115] were also given to identify and compensate an unknown deadzone.

In the above studies, the input deadzone was assumed to be independent of the system operating conditions. Such an assumption does not hold when we deal with a control process with a high-precision requirement, say controlling an actuator at the micrometer or nanometer scale. Figure 4.1 shows the values of an input deadzone physically measured from a linear piezoelectric motor. The horizontal axis is the motor position, and the vertical axis is the minimum input voltage needed for the motor to overcome its deadzone. "∗" indicates the left bound of the deadzone and "+" indicates the right bound of the deadzone. It can be seen that the deadzone is not symmetric, and its size is dependent on the displacement of the motor position. Further investigation shows that the deadzone size also varies, though not severely, according to a number of other factors such as the room temperature. In a word, we have to deal with a deadzone that is non-linear, time-varying, unknown, state dependent, *etc.*

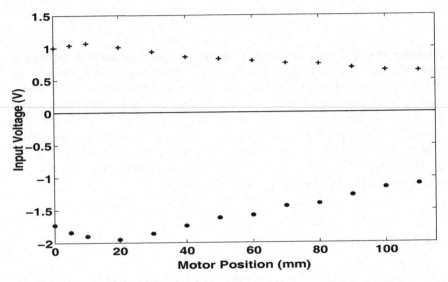

Fig. 4.1 Deadzone vs. motor position displacement

Recently, we disclosed a new finding [159] that ILC algorithms, originally designed for systems without input deadzone, can effectively compensate the nonlinear deadzone through control repetitions. The ILC algorithms can automatically generate appropriate control profiles to achieve the deadzone inverse for the given control process. To support the new finding, the simplest ILC algorithm that is designed originally for systems without input deadzone, was considered. In a rigorous mathematical manner, it was proven that, despite the presence of the input deadzone, the simplest ILC algorithm retains its ability of achieving satisfactory performance in tracking control.

The significance of the new finding is two-fold. First, this finding implies that most existing ILC algorithms developed in the past 20 years can be directly applied to systems either with or without input deadzone. Second, we have a new approach to deal with the input deadzone, in particular to deal with the non-linear and state-dependent deadzone for which existing deadzone control algorithms are unable to effectively compensate.

With the theoretical results in [159], in this chapter we first present a general model of an input deadzone that is non-linear in states, and discuss the class of dynamic processes that can be controlled under ILC while the input deadzone is unknown. Next, as a demonstration we present a real-time implementation of a simple ILC on a piezoelectric motor with a position-related input deadzone. The experimental result shows precise tracking performance of 1 micrometer, while the deadzone could reach 20% of the range of the control input.

The input saturation is another input non-linearity commonly encountered in real-time control. The input saturation is in fact a class of input constraints. From an engineering point of view, most physical actuators, sensors and interfacing devices such as AD/DA, are subject to saturation because of the existence of hardware limitations. For instance most actuators including electrical, mechanical, hydraulic and pneumatic ones, have limited output. A motor can only generate a limited torque, an electrical drive circuit can only produce a limited current, which can be modelled mathematically by a saturation. Also, in real-world applications often a soft limiter is added to a control system, with the purpose of preventing over-large control actions that may be harmful to the control apparatus.

The saturation phenomenon has been analyzed in the field of "anti-windup" for linear time-invariant system [32, 47, 134], in which an additional feedback is introduced in such a way that the actuator could stay in the linear part, *i.e.* unsaturated part so that the problems caused by saturation could be avoided. However, this technique requires the complete knowledge even for linear time-invariant systems, hence it is difficult to apply to ILC problems that deal with highly non-linear and uncertain systems.

On the other hand, input saturation has been well investigated in the ILC literature [33, 55, 122, 156]. It has been made clear that the ultimate convergence of ILC is not affected by the input saturation when the tracking task is a feasible one under the input constraints. In this chapter we will present a general model of an input saturation, and briefly discuss the ILC properties when the tracking task is no longer a feasible one.

Input backlash is also a kind of non-smooth and non-affine-in-input factor widely existing in actuators. Comparing with input deadzone and input saturation, input backlash is more difficult to handle because input deadzone and input saturation are memoryless, whereas input backlash has an element of memory.

System control with backlash has been studied since 1940. In [27, 126], based on online identification of backlash parameters, the adaptive inverse compensation methods were proposed. Switched control [88], dithered control [29], and Taylor's SIDF method [133] have also been applied to reduce or eliminate influence from backlash. Owing to the capability of learning any non-linear functions, neural networks have been used to identify and compensate backlash. A recurrent neural network with unsupervised learning by a genetic algorithm was developed in [117]. In [114], a neural network was proposed to handle gear backlash in precision-position mechanisms. Due to the asymptotic convergence property, these methods are not applicable to control tasks that end in a finite time interval.

However, it is still an open problem to rigorously prove ILC convergence when input backlash exists. In this chapter we will present a general model of input backlash and some simulation results with ILC.

This chapter is organized as follows. In Sect. 4.2, the control problem associated with a class of discrete-time non-linear systems and a non-linear state-dependent deadzone is presented. In Sect. 4.3, the control problem associated with an input saturation is mathematically formulated and discussed. In Sect. 4.4, the characteristics of a kind of backlash are formulated and the ILC performance is demonstrated through a numerical example. Section 4.5 exhibits an application example of ILC to a piezoelectric motor. Finally, some onclusions are drawn in Sect. 4.6.

4.2 ILC with Input Deadzone

Consider the following discrete-time dynamic system

$$\begin{aligned}
\mathbf{x}_i(k+1) &= \mathbf{f}(\mathbf{x}_i(k),k) + \mathbf{b}(\mathbf{x}_i(k),k)u_i(k) \\
u_i(k) &= DZ[v_i(k)] \\
y_i(k+1) &= \mathbf{c}\mathbf{x}_i(k+1),
\end{aligned} \tag{4.1}$$

where $\mathbf{x}_i \in \mathscr{R}^n$ is the system state, $y_i \in \mathscr{R}$ is the physically accessible system output, $u_i \in \mathscr{R}$ is the plan input, but not available for control, $v_i \in \mathscr{R}$ is the actual system input, $\mathbf{f}$ and $\mathbf{b}$ are non-linear time-varying vectors, $\mathbf{c}$ is a constant vector. $DZ[\cdot]$ represents the input deadzone and the characteristic is defined as

$$u_i(k) = DZ[v_i(k)] = \begin{cases} m_{r,i}(k)[v_i(k) - \eta_{r,i}(k)] & v_i(k) \in I_{r,i}(k) \\ 0 & v_i(k) \in I_{D,i}(k) \\ m_{l,i}(k)[v_i(k) - \eta_{l,i}(k)] & v_i(k) \in I_{l,i}(k) \end{cases} \tag{4.2}$$

where $m_{l,i}(k) > 0$ and $m_{r,i}(k) > 0$ are slopes, $\eta_{l,i}(k) \leq 0$ and $\eta_{r,i}(k) \geq 0$ are the left and right boundary points of a deadzone. $I_{D,i} \triangleq [\eta_{l,i}(k), \eta_{r,i}(k)]$ is the region of deadzone. $I_{l,i} \triangleq (-\infty, \eta_{l,i}(k)]$ is the left region of the deadzone. $I_{r,i} \triangleq [\eta_{r,i}(k), +\infty)$ is the right region of the deadzone.

The deadzone at an operating point $(\mathbf{x}(k),k)$ is shown in Fig. 4.2. Note that the deadzone can be non-symmetric, and both its width and slopes can be time varying and state dependent.

Note that the formulated control problem in fact has two dimensions. One is along the time sequence k, which belongs to a finite time interval, and the dynamical system is always repeated over the interval. The other one is the iteration axis denoted by i that approaches infinity.

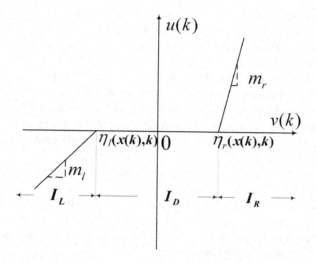

Fig. 4.2 The deadzone non-linearities $u(k) = DZ[v(k)]$

The dynamical system (4.1) and the deadzone (4.2) are assumed to satisfy the following assumptions.

Assumption 1 $f_i(k)$, $b_i(k)$, $\eta_{l,i}(k)$, $\eta_{r,i}(k)$, $m_{l,i}(k)$, and $m_{r,i}(k)$ are Global Lipschitz continuous (GLC) functions with respect to the argument $x_i(k)$, and associated with unknown Lipschitz constants l_f, l_b, $l_{\eta,l}$, $l_{\eta,r}$, $l_{m,l}$, and $l_{m,r}$ respectively. In addition, $m_{l,i}(k)$ and $m_{r,i}(k)$ are upperbounded.

Assumption 2 System (4.1) satisfies the identical initial condition (i.i.c.): $\forall i$, $\delta x_i(0) \overset{\triangle}{=} x_d(0) - x_i(0) = 0$. Hence $e_i(0) \overset{\triangle}{=} y_d(0) - y_i(0) = 0$.

Assumption 3 $cb_i(k) > 0$ with known upperbound.

Based on contraction mapping, all existing ILC methods require Assumption 1. The reason is as follows. ILC methodology tries to use as little system prior knowledge as possible in its design, for instance it does not use any information regarding the state dynamics that may not be available in many practical control problems. The lack of such system knowledge, however, gives rise to a difficulty in designing a suitable (stable) closed-loop controller. Hence, ILC often works in open loop in the time domain but closed loop in the iteration domain. The GLC condition warrants no finite escape-time phenomenon even if the open-loop system is unstable. In practice, many real processes have finite operating rangesdue to physical limits, hence can be treated essentially as GLC even if the dynamics are highly non-linear. In fact, if the system dynamical information is available and used, then by means of Lyapunov methods ILC can be extended to local Lipschitzian functions, as stated in Chap. 2. In this chapter, our main objective is to cope with the non-linear input singularity problem arising from the deadzone, hence we consider GLC so that our analysis and conclusion are applicable to most existing ILC schemes.

As discussed in Chap. 2, ILC tries to make a "perfect" tracking in a finite time interval, whereas other control methods aim at an asymptotical convergence in an infinite time interval. A perfect tracking from the beginning demands the perfect initial condition, that is, Assumption 2. Many practical control problems require such a perfect tracking over the entire transient period including the initial one, such as the track following in a hard disk drive, or temperature control in a wafer process. In fact, in order to fulfill such a control task, other control methods will also need this assumption.

Assumption 3, *i.e.* $\mathbf{cb}_i(k) > 0$, gives the controllability condition imposed on non-linear systems.

The ultimate control target is to find an appropriate control input sequence $v_i(k)$ that uses only output signals, such that the output sequence $y_i(k)$ converges to the desired output $y_r(k)$ iteratively as $i \rightarrow \infty$. Here, $y_r(k)$ can be regarded as the output generated by the following dynamics

$$\mathbf{x}_r(k+1) = \mathbf{f}_r(k) + \mathbf{b}_r(k)u_r(k)$$
$$y_r(k) = \mathbf{cx}_r(k), \tag{4.3}$$

where $\mathbf{f}_r(k) \triangleq \mathbf{f}(\mathbf{x}_r, t)$, $\mathbf{b}_r(k) \triangleq \mathbf{b}(\mathbf{x}_r, k)$ and $u_r(k) = DZ[v_r(k)]$ with $u_r(k)$ the ideal control input without deadzone and $v_r(k)$ the actually desired control input.

The simplest ILC algorithm in discrete time is

$$v_{i+1}(k) = v_i(k) + \beta e_i(k+1) \tag{4.4}$$
$$e_i(k) = y_r(k) - y_i(k),$$

where $\beta > 0$ is the learning gain chosen to satisfy the condition

$$|1 - \beta \mathbf{cb}| \leq \rho < 1. \tag{4.5}$$

It can be seen that (4.4) is a default ILC algorithm, and the condition (4.5) in general holds for processes without deadzone, namely $DZ[\cdot] = 1$ globally in (4.1). In the presence of input deadzone, it is difficult to directly derive the convergence condition (4.5). Instead, in [159] it proves that the simplest ILC algorithm (4.4), when directly applying to the process (4.1), achieves a point-wise convergence

$$u_i(k) \rightarrow u_r(k) \qquad k = 1, 2, \cdots, N$$

under the *i.i.c.* condition. In Sect. 4.5, we will conduct an experiment to show its effectiveness.

4.3 ILC with Input Saturation

The input with saturation can be modelled as

$$u_i(k) = SAT[v_i(k)],$$

where $v_i(k)$ is updated according to (4.4) The characteristic of the saturation is defined as

$$SAT[v_i(k)] = \begin{cases} u_{\max}(k) & v_i(k) \geq u_{\max}(k) \\ v_i(k) & v_{\min}(k) < v_i(k) < u_{\max}(k) \\ u_{\min}(k) & v_i(k) \leq u_{\min}(k) \end{cases}, \qquad (4.6)$$

where $u_{\min}(k)$ and $u_{\max}(k)$ are input constraints. Note that the saturation can be non-symmetric and the quantities $u_{max}(k)$ and $u_{min}(k)$ can be positive or negative.

The effect of a saturation with regard to a control task depends on whether the task is feasible under the input constraints. First, consider the scenario where a task is feasible under the input constraints. Most ILC algorithms with input constraints were assumed to satisfy this condition, which is summarized by the following assumption.

Assumption 4 *For a given control task $y_r(k)$, the ideal control input profile $u_r(k)$ satisfies the feasibility condition*

$$u_{\min}(k) \leq u_r(k) \leq u_{\max}(k).$$

Under this assumption, it has been proven that the ILC can guarantee the learning convergence in the iteration domain. This conclusion was obtained for ILC when analyzed using either contractive mapping [55] or the Lyapunov method [156]. In such circumstances, there is no need to take input constraints into account when designing an ILC algorithm.

Next, consider the scenario where a task is infeasible under the input constraints. It should be pointed out that such a scenario reflects an improper task planning or scheduling for $y_r(k)$. Hence it is in general a re-planning problem instead of a control problem. What is interested to look at is whether ILC can still perform reasonably well under such a tough scenario where the task assigned is never achievable. For simplicity, we consider the case $u_r(k) > u_{\max}(k)$ for a given $y_r(k)$. The other case with $u_r(k) < u_{\min}(k)$ can be discussed in a similar way.

We will show that, even if $y_r(k)$ is not attainable due to insufficient control capacity, the ILC will still push $y_i(k)$ close to $y_r(k)$. This interesting quality consists of two properties of ILC: 1) the output $y_{\max}(k)$, generated by $u_{\max}(k)$, is the closest to $y_r(k)$; 2) ILC input $u_i(k)$ will converge to the correct limit $u_{\max}(k)$.

The first property can be easily derived with a mild assumption that the larger the process input, the larger the process output. Most physical systems or engineering systems do satisfy this assumption. For instance, the larger the supply current, the larger the torque generated by an electrical motor. Similarly, the higher the ramp inflow entering the freeway, the higher the resulting freeway density [46]. In other words, the higher is $u_i(k)$, the higher is $y_i(k)$. Thus, when $u_i(k)$ reaches $u_{max}(k)$, the corresponding $y_i(k)$ should reach the maximum value, denoted as $y_{max}(k)$. Since $u_r(k) > u_{max}(k)$ at the instant k, we have $y_r(k) > y_{max}(k)$. Note that $u_{max}(k)$ is always larger than the unsaturated $u_i(k)$, thus $y_{max}(k)$ is the closest to $y_r(k)$ comparing with $y_i(k)$ which is generated by any unsaturated input $u_i(k)$.

Now derive the second property. Note that the relationship $y_r(k) > y_{max}(k) \geq y_i(k)$ holds for all iterations, $e_i(k) = y_r(k) - y_i(k) > 0$. From the learning control algorithm (4.4), it can be seen that ILC is a discrete integrator with a positive input $\beta e_i(k+1)$. Hence, with a finite number of iterations the control input $u_{i+1}(k)$ will reach its upper limit $u_{max}(k)$.

4.4 ILC with Input Backlash

Input backlash shares some similarities with input deadzone but is a more general phenomenon. Consider a class of backlash described by the following equation

$$u_i(k) = BL[v_i(k)] = \begin{cases} m_l[v_i(k) - \eta_l] & v(k) \in I_{L,i}(k-1) \\ u_i(k-1) & v(k) \in I_{D,i}(k-1) \\ m_r[v_i(k) - \eta_r] & v(k) \in I_{R,i}(k-1) \end{cases} . \tag{4.7}$$

where $m_l > 0, m_r > 0, \eta_l \leq 0, \eta_r \geq 0, k \in \mathscr{K}, i \in \mathscr{Z}_+; I_{L,i}(k-1) \triangleq (-\infty, v_{l,i}(k-1))$, $I_{D,i}(k-1) \triangleq [v_{l,i}(k-1), v_{r,i}(k-1)]$ and $I_{R,i}(k-1) \triangleq (v_{r,i}(k-1), \infty)$ with $v_{l,i}(k-1) = \frac{u_i(k-1)}{m_l} + \eta_l$ and $v_{r,i}(k-1) = \frac{u_i(k-1)}{m_r} + \eta_r$.

The characteristic of the backlash can be described as Fig. 4.3. Unlike input deadzone which is memoryless or a static mapping between u_i and v_i, input backlash has an element of memory because the current controller output $u_i(k)$ depends not only on current control input $v_i(k)$, but also the past controller output $u_{i-1}(k)$ as shown in the figure and preceding model (4.7) where the boundaries of three regions, $v_{r,i}$ and $v_{l,i}$, are relevant to $u_{i-1}(k)$.

The question is whether the simplest ILC algorithm (4.4) can let the output $y_i(k)$ of the process (4.1) track the output $y_r(k)$ of the reference model (4.3). This is still an open problem in the field of ILC, though it has been partially solved in the field of adaptive control [126].

To illustrate the effectiveness of ILC, consider the following sampled-data system

$$x_1(kT_s + T_s) = x_2(kT_s)$$
$$x_2(kT_s + T_s) = 0.4sin[x_1(kT_s)] + 0.15x_2(kT_s) + BL[v(kT_s)]$$
$$y(kT_s + T_s) = x_2(kT_s + T_s),$$

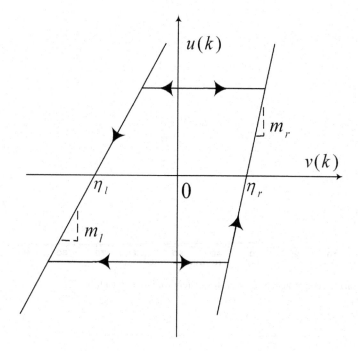

Fig. 4.3 The backlash non-linearities $u(k) = BL[v(k)]$

where the backlash parameters are $\eta_l = -1.3$, $\eta_r = 1.5$, $m_l = 1.1$ and $m_r = 1.0$. The desired output is $y_r(k) = 10sin^3(kT_s)$, $k = \{0, 1, \cdots, 6283\}$. The initial states are $x_{2,i}(0) = y_r(0) = 0$ and $x_{1,i} = 0$, satisfying the *i.i.c.* condition.

Assume the known bound of m_l, m_r and **cb** are $B_1 = 1.2$ and $B_2 = 1.2$ respectively. Choose $\beta = 0.6$ to guarantee $0 < 1 - \beta B_1 B_2 < 1$. Let the sampling period $T_s = 0.001$ s. By applying the simplest law (4.4), the simulation result is shown in Fig. 4.4. The horizon is the iteration number and the vertical is $|y_r - y_i|_{sup}$. Figure 4.5 shows the control signal v_i at the $100th$ iteration.

However, more efforts are needed to verify the effectiveness of ILC through theoretical analysis and real-time experiments.

4.5 ILC Implementation on Piezoelectric Motor with Input Deadzone

To verify the effectiveness of learning control in the presence of an unknown nonlinear deadzone, experiments have been carried out using a linear piezoelectric motor that has many promising applications in industries. The piezoelectric motors are characterized by low speed and high torque, which are in contrast to the

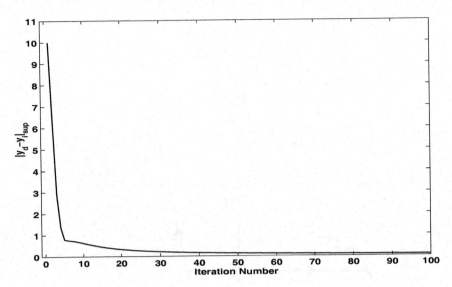

Fig. 4.4 Learning convergence of $y_r - y_i$ for system with input backlash

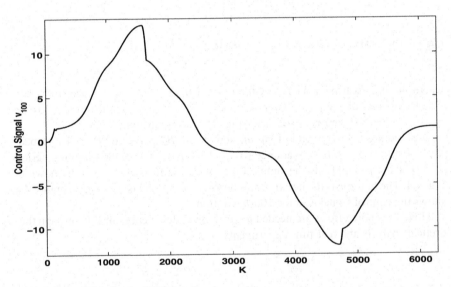

Fig. 4.5 Control signal at the 100th iteration

high-speed and low-torque properties of the conventional electromagnetic motors. Moreover, piezoelectric motors are compact, light, operating quietly, and robust to external magnetic or radioactive fields. Piezoelectric motors are mainly applied to high-precision control problems as they can easily reach the precision scale of micrometers or even nanometers. This gives rise to an extra difficulty in establishing an accurate mathematical model for piezoelectric motors: any tiny factors, non-linear and unknown, will severely affect their characteristics and control performance. In particular, piezoelectric motors have a huge deadzone in comparison with their admissible level of input signals, and the size of deadzone varies with respect to the position displacement. Figure 4.1 shows the characteristics of the deadzone pertaining to the piezoelectric motor used in this experiment.

The configuration of the whole control system is outlined in Fig. 4.6. The driver

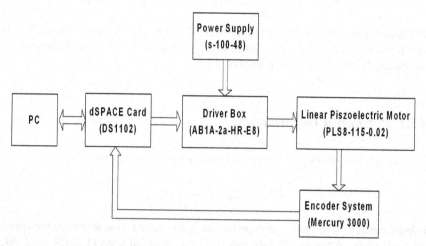

Fig. 4.6 Structure of the control system

and the motor can be modelled approximately as

$$\dot{x}_1(t) = x_2(t)$$
$$\dot{x}_2(t) = -\frac{k_{fv}}{M}x_2(t) + \frac{k_f}{M}u(t)$$
$$y(t) = x_1(t), \tag{4.8}$$

where x_1 is the motion position, x_2 is the motion velocity, $M = 1$ kg is the moving mass, $k_{fv} = 144$ N is the velocity damping factor and $K_f = 6$ N/V is the force constant.

The machine sampling time is $T_s = 0.004$ s. The discretized model of (4.8) is

$$x_1(k+1) = x_1(k) + 0.003x_2(k) + (6.662 \times 10^{-6}u(k))$$
$$x_2(k+1) = 0.5621x_2(k) + 0.003u(k)$$

$$y(k+1) = x_1(k+1). \tag{4.9}$$

The simple linear model (4.9) does not contain any non-linear and uncertain effects such as the frictional force in the mechanical part, high-order electrical dynamics of the driver, loading condition, *etc.*, which are hard to model in practice. In general, producing a high-precision model will require more effort than performing a control task with the same level of precision. Note that the piezoelectric motor deadzone is not only non-symmetric but also affected by the motor position.

Let $T = 4$ s, hence $k \in \{0, 1, \cdots, 1000\}$. According to the specification provided by the manufacturer, the piezoelectric motor possesses a system repeatability of 0.1 μm.

Consider the tracking problem with the desired target

$$y_d(k) = [20 + 60\sin(0.35kT_s)]\text{mm}, \quad k \in \{0, \cdots, 1000\}.$$

The system initial position for all iterations is set to be $x_1(0) = 20$ mm and $x_2(0) = 0$, which is realized by a PI controller in the experiments. The *i.i.c.* in Assumption 2 can be satisfied to the precision level of 0.1 μm.

4.5.1 PI Control Performance

First, let us examine the control performance when a PI controller

$$U(z) = (k_p + k_I \frac{T_s}{z-1})E(z).$$

is implemented. A sequence of experiments is conducted so as to decide the appropriate PI gains. When the PI gain pair (k_p, k_I) is doubled from $(1.5, 15)$ to $(3, 30)$, the tracking error has been reduced by about half (Fig. 4.7). However, when (k_p, k_I) is further increased from $(3, 30)$ to $(6, 60)$, the maximum tracking error remains almost the same but chattering appears in the control signal and motor response, which is very harmful to such a precision device. Thus, the PI gains are chosen to be $(k_p, k_I) = (3, 30)$. Note that the initial phase of large tracking error is due to the lack of control action in the deadzone.

4.5.2 ILC Performance

The control and error profiles obtained from the PI control with $(k_p, k_I) = (3, 30)$ are used as the initial values for $v_0(k)$ and $e_0(k)$. Since repeating the same PI control will only reproduce the same tracking performance, we introduce a learning mechanism aiming at a better response. In real-time applications, a pure open-loop ILC, namely PCL, is not recommended because of the lack of robustness in the presence of non-

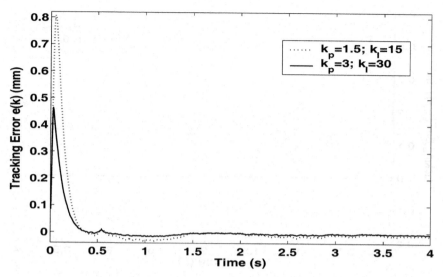

Fig. 4.7 Tracking errors with different k_p and k_I

repeatable components such as the measurement noise. In our experiment, a simple proportional controller working as the current cycle controller is combined with the ILC scheme and the control law can be described as:

$$u_{i+1}(k) = u_i(k) + \beta_l e_i(k+1) + \beta e_{i+1}(k).$$

The feedback part can be treated as a part of the system dynamics $\mathbf{f}(\mathbf{x}(k), k)$, hence the previously derived learning convergence property still holds. Clearly the ILC is PCCL.

The learning gain is chosen to be $\beta_l = 0.6$ and the feedback gain $\beta = 3$. It is easy to verify that the learning convergence condition is satisfied with the chosen β. The tracking errors for iterations $i = 0$ which is PI control, $i = 1$, and $i = 20$ are given in Fig. 4.8. In order to clearly exhibit and compare the tracking performance among PI and ILC at different iterations, we split the tracking interval into two subintervals $[0, 0.4]$ s and $(0.4, 4]$ s. Both PI and ILC need to work over the first interval $\mathscr{K}_1 \triangleq \{0, \cdots, 100\}$ to overcome the influence from the deadzone. The system enters the steady state in the second interval $\mathscr{K}_2 \triangleq \{101, \cdots, 1000\}$. The tracking error and control signal profiles of PI, 1st iteration and 20th iteration over the first interval are shown in Fig. 4.9 and Fig. 4.10, respectively. It is clearly shown that, based on the iterative updating, the system input is able to move out of the input deadzone earlier and the tracking error is reduced accordingly. The maximum tracking error is reduced from 0.45 mm with the PI control to 0.1 mm after 20 iterations, a reduction rate of approximately 97%. The effect of learning is immediately obvious.

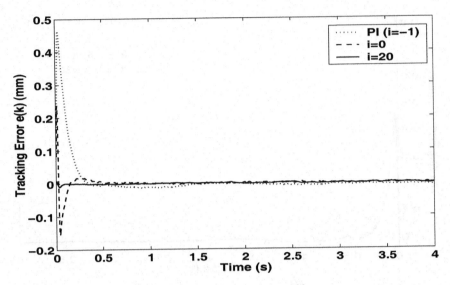

Fig. 4.8 Comparison of tracking errors at iterations $i = 0$, 1 and 20

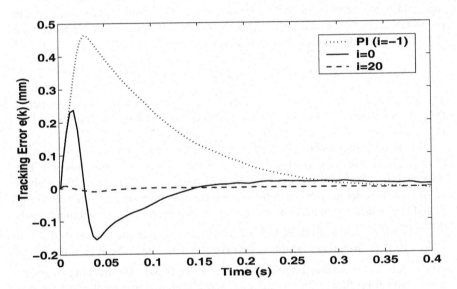

Fig. 4.9 The tracking errors at different iterations ($k \in \mathscr{K}_1$)

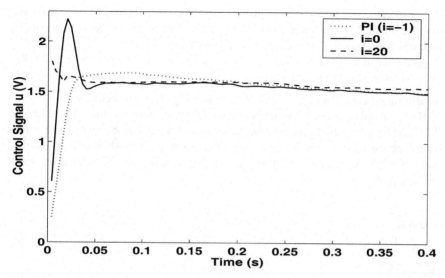

Fig. 4.10 The control signals at different iterations ($k \in \mathcal{K}_1$)

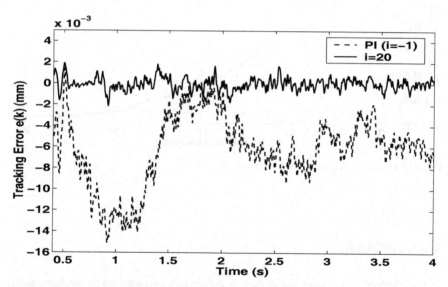

Fig. 4.11 Comparison of tracking errors at iterations $i = 0$ and 20 ($k \in \mathcal{K}_2$)

The steady-state tracking error profiles for the PI ($i = 0$) and ILC ($i = 20$) over the time interval $\mathscr{K}_2$ are given in Fig. 4.11. ILC is able to achieve a precision of ± 2 μm, which is about 7 times better than that of PI control.

To demonstrate the whole learning process, the maximum tracking error in $\mathscr{K}_1$, *i.e.* $\max_{k \in \mathscr{K}_1} |e(k)|$, and the maximum steady state tracking error, *i.e.* $\max_{k \in \mathscr{K}_2} |e(k)|$, of each iteration are recorded and given in Fig. 4.12 and Fig. 4.13, respectively.

Here, we only gave the experimental results for a tracking problem. The simplest ILC can also be applied to regulation or set-point tracking problems and the better control performance can be achieved. The experimental results show that the steady-state error can be reduced to 1 μm after several learning iterations for a set-point tracking problem.

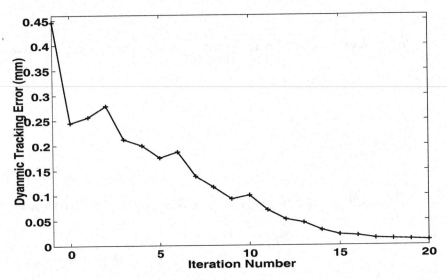

Fig. 4.12 Learning convergence in the interval $\mathscr{K}_1$

4.6 Conclusion

In this chapter, three non-smooth factors commonly encountered in mechanical or mechatronics systems were discussed. By formulating these non-smooth factors mathematically, we can prove the learning convergence when ILC is applied. We show that basic ILC algorithms, originally designed for plants without such non-smooth factors, are able to automatically compensate these non-smooth factors. As a consequence, all existing ILC schemes based on contraction mapping can be used straightforwardly for systems with input deadzone, input backlash and input saturation. This fact is important as in practice a control-system designer is unlikely to

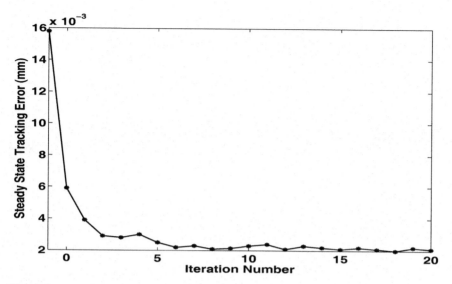

Fig. 4.13 Learning convergence in the interval $\mathscr{K}_2$

have any accurate models for non-smooth factors, or even be unaware of the presence of non-smooth factors. This novel property of ILC is further validated through real-time experiments on a precision servo. It is worth pointing out that, at least for repeated control systems, we can now handle these non-smooth factors that are non-linear, unknown, time varying and state dependent, which could be difficult for other compensation methods to cope with.

Future works include the extension to MIMO systems, the extension to state control and to dynamical output control, the presence of two or more non-smooth non-linearities in input, the presence of random variations in the non-smooth non-linearities.

Chapter 5
ILC for Process Temperature Control: Application to a Water-heating Plant

Abstract ILC for enhancing tracking in batch processes, which can be approximated by a first-order plus dead-time (FOPDT) model, is presented. Enhancement is achieved through filter-based iterative learning control. The design of the ILC parameters is conducted in the frequency domain, which guarantees the convergence property in the iteration domain. The filter-based ILC can be easily added on to existing control systems. To clearly demonstrate the features of the proposed ILC, a water-heating process under a PI controller is used as a testbed. The empirical results show improved tracking performance with iterative learning.[1]

5.1 Introduction

Trajectory tracking, whose primary control target is to track a specified profile as tightly as possible in a finite time interval, is very common in process-control problems, such as concentration control of a chemical reactor in pharmaceutic industry, or temperature control in the wafer industry. In practice, the most widely used control schemes in process-control industries are still PI or PID controllers with modifications, owing to the simplicity, easy tuning, and satisfactory performance [116, 53, 146]. On the other hand, many advanced control schemes have been proposed to handle complicated control problems. Nevertheless, it is still a challenging control problem where perfect trajectory tracking is concerned, *i.e.* how to achieve satisfactory tracking performance when the process is under transient motion over the entire operation period. Most advanced control schemes can only achieve perfect tracking asymptotically – the initial tracking will be conspicuously poor within the finite interval. PI or PID control schemes in most cases can only warrant a zero steady-state error.

---

[1] With Elsevier permission to re-use "Enhancing trajectory tracking for a class of process control problems using iterative learning," coauthored by Xu, J.-X., Lee, T.H., Tan, Y., Engineering Applications of Artificial Intelligence, The International Journal of Intelligent Real-Time Automation, Vol. 15, pp.53–64, 2002.

There are many industrial processes under batch operations, which by their virtue are repeated many times with the same desired tracking profile. The same tracking performance will thus be observed, albeit with hindsight from previous operations. Clearly, these continual repetitions make it conceivable to improve tracking, potentially over the entire task duration, by using information from past operations. To enhance tracking in repeated operations, ILC schemes developed hitherto well cater to the needs. Numeric processing on the acquired signals from previous iterations yields a kind of new feed-forward compensation, which differs from most existing feedforward compensations that are highly model based. Comparing with many feedforward compensation schemes, ILC requirements are minimal – a memory storage for past data plus some simple data operations to derive the feed-forward signal. With its utmost simplicity, ILC can very easily be added on top of existing (predominantly PID batch) facilities without any problems at all.

In this chapter, ILC is employed to enhance the performance of a kind of process dynamics, which can be characterized more or less by the first-order plus dead time (FOPDT) model. The approximated model is usually obtained from the empirical results. It has been shown that this approximation model, though very simple, has been used for near sixty years and is still widely adopted [113]. Based upon this FOPDT, the famous Ziegler–Nichols tuning method [164] was developed and nowadays has become an indispensable part of control textbooks [94].

However, when higher tracking performance is required, feedback and feedforward compensations based on the FOPDT model may not be sufficient due to the limited modelling accuracy. In such circumstances, ILC provides a unique alternative: to reconstruct and capture the desired control profile iteratively through past control actions, as far as the process is repeatable over the finite time interval. In this chapter, the filter-based learning control scheme is incorporated with PI control in order to improve the transient performance in the time domain.

The filter-based ILC scheme is proven to converge to the desired control input in the frequency domain within the bandwidth of interest. The bandwidth of interest can be easily estimated using the approximated FOPDT model. The proposed ILC scheme simply involves 2 parameters – the filter length and the learning gain, both can be easily tuned using the approximated model. Also, this scheme is practically robust to random system noise owing to its non-causal zero-phase filtering nature. A water-heating plant is employed as a testbed to illustrate the effectiveness of the proposed filter-based learning scheme.

The chapter is organized as follows. Section 5.2 formulates the control problem of FOPDT in general, and the modelling of a water-heating plant in particular. Section 5.3 gives an overview of filter-based ILC with its convergence analysis in the frequency domain. Section 5.4 details the controller design work and the experimental results. From these results, a modified ILC scheme with profile segmentation and feedforward initialization, is used to improve tracking performance even further. Finally, Sect. 5 concludes the chapter.

5.2 Modelling the Water-heating Plant

The first-order plus dead-time model, FOPDT, has been widely used in industries to approximate various kinds of type-0 processes

$$P(s) = \frac{Ke^{-\tau s}}{1 + sT_a} \tag{5.1}$$

where K is the plant steady-state gain, T_a is the apparent time constant and τ is the apparent dead-time. There are two main reasons accounting for the popularity of the FOPDT model. The first is its extremely simple structure associated with only three parameters that can be easily calculated through empirical tests, for example a simple open-loop step response [94]. The second is the well-developed PI and PID auto-tuning rules that provide a simple and effective way for setting controller parameters [164, 39].

The FOPDT model provides a very convenient way to facilitate the PI/PID controller setting, regardless of the existence of non-linear factors, higher-order dynamics, or even distributed parametric dynamics, *etc*. In practice, the FOPDT-model-based PI/PID control has been effectively applied to two kinds of control problems. One is the regulation problem where the objective is to maintain the operating point, and another is the set-point control problem where the objective is to achieve a pre-specified step response with balanced performance among settling time, overshoot and rising time. However, this simple control strategy, which is based on a simple model approximation such as FOPDT, and simple control schemes such as PID, may not be suitable for more complicated control tasks with high tracking precision requirement. Consider a control target: tracking a temperature profile as shown in Fig. 5.1. This is the simplest trajectory consisting of two segments: a ramp and a

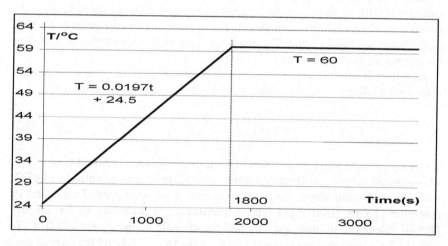

Fig. 5.1 Desired temperature profile

level, starting at $t = 0$ and ending at $t = 3600$ s.

A few problems arise when such a trajectory tacking task is to be fulfilled. First, a single integrator is inadequate to follow a ramp signal. Adding more integrators will degrade the system performance due to the extra phase lag. Second, even if the controller is able to follow the reference in the steady state, the transient performance will be poor, because of the finite tracking period. Third, since only an approximate FOPDT model is available, many advanced control schemes, which require much of the modelling knowledge, cannot be applied. Accurate modelling, on the other hand, is time consuming, costly, and usually a tougher problem than control.

It is worth pointing out, when a control task is performed repeatedly, we gain extra information from a new source: past control input and tracking error profiles. Different from most control strategies, iterative learning control explores the possibility of fully utilizing this kind of system information, subsequetly enhancing tracking for the next run. An immediate advantage is that the need for process model knowledge can be significantly reduced. We will show later that, in order to design an ILC in the frequency domain based on the simple FOPDT model, it is sufficient to know approximately the bandwidth of the desired control profile. Another advantage is that we can retain the auto-tuned or heuristically tuned PI/PID, and simply add the ILC mechanism. In this way the characteristics of the existing feedback controller can be well reserved, and ILC provides extra feedforward compensation.

To facilitate the discussions on modelling and control in general, and verify the effectiveness of ILC with FOPDT in particular, a water-heating process is used in this work. Figure 5.2 is a schematic diagram showing the relevant portions of the water-heating plant. Water from tank A is pumped (N1) through a heat exchanger as the cooling stream. The heating stream on the other side of the exchanger is supplied from a heated reservoir. This heated stream is pumped (N2) through the exchanger before returning to the reservoir where it is heated by a heating rod (PWR). A distributed model based on physical principles gives a first-order partial differential equation (PDE), as shown in the Appendix. Figure 5.3 shows the responses obtained from the experiment with the actual plant, and the simulation with the PDE model, to a step input of 200 W at the heater (the heater is switched to 200 W at $t = 0$ s and then switched off at $t = 12500$ s).

Both the simulation and the experimental responses show that the water-heating plant can be effectively approximated by the following FOPDT system

$$P(s) = \frac{T2(s)}{PWR(s)} = \frac{0.1390}{2865s + 1}e^{-50s}, \tag{5.2}$$

where $T2(s)$ is the water temperature in the heated reservoir. The dead-time $\tau = 50$ s is measured in the experiment. Since the process can be approximated by FOPDT, normally a PI controller is designed to realize a set-point control task. The PI controller can be tuned using relay experiments [7] according to the Ziegler-Nichols rules. The ultimate gain K_u and the ultimate period P_u are 125.0 $°C/W$ and 512.2 s, giving the proportional gain $k_p = 0.45K_u = 56.8$ and the integral gain $k_I = 0.54P_u/K_u = 0.1318$. Now, let the target trajectory be the piecewise-smooth

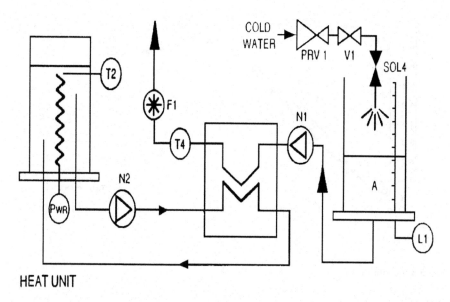

HEAT UNIT

Fig. 5.2 Schematic diagram of a water heating plant

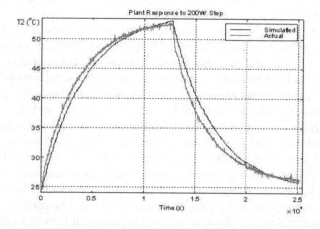

Fig. 5.3 Model and plant response to a 200 W step

temperature profile shown in Fig. 5.1. The control objective is for T2 to track the trajectory as closely as possible over its entire duration, by varying the input to the heater (PWR). Both pumps (N1 and N2) are maintained at pre-set values throughout the run. The tracking response under PI control is shown in Fig. 5.4.

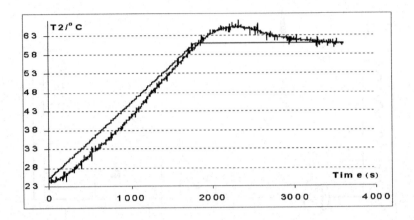

Fig. 5.4 T_2 with PI control ($k_p = 56.8$, $k_I = 0.1318$)

From the experimental result we can observe the large tracking discrepancy at the initial stages of the ramp and level segments. The control system is not able to make a quick response to the variations of the target trajectory, due to the inherent phase lag nature of the process and the feedback controller. Such a phase lag could have a strong influence on the transient performance. The most effective way to overcome the difficulty incurred by the system delay is to incorporate a feedforward compensation, which in the ideal case should be able to provide the desired control input profile $u_r(t)$. For the water-heating process, the estimated control input profile $\hat{u}_r(t)$, which is an approximation of $u_r(t)$, can be calculated according to the approximate FOPDT model (5.2) and the target trajectory given by Fig. 5.1, as below.

Theoretically speaking, if the FOPDT model (5.2) is accurate, we can use $\hat{u}_r(t)$ directly to produce the desired temperature profile. In practice, however, the model cannot be very precise due to the nature of approximating a PDE process by a simple FOPDT, as well as numerous imperfect factors such as measurement noise, disturbances, *etc*. For instance, if the steady-state gain estimated is with 10% deviation, an off-set will be produced in $\hat{u}_r(t)$ also around 10%. This is a fundamental limitation of all kinds of model-based feedforward compensations. ILC, which updates feedforward compensation based on the past control input and tracking error profiles, avoids such problems.

On observing Fig. 5.5, we also note that $\hat{u}_r(t)$ is not implementable in practice because of the presence of the discontinuity, which demands a theoretically infinite bandwidth in actuation if a perfect tracking is required. A piecewise-continuous trajectory can be commonly found in process control, *e.g.* the desired temperature profile, the desired concentration profile, the desired pressure profile, *etc*. From the spectrum of $\hat{u}_r(t)$ (Fig. 5.6), however, it shows that the frequency content of the estimated control signal $\hat{u}_r(t)$ is almost negligible for $\omega/2\pi > 0.003$ Hz with the task duration $T = 3600$ s. Thus, a filter-based ILC will be able to learn the desired control signals within a finite bandwidth.

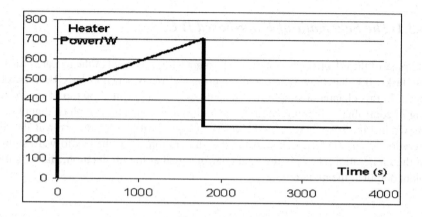

Fig. 5.5 Estimated control input $\hat{u}_r(t)$ based on the approximated FOPDT

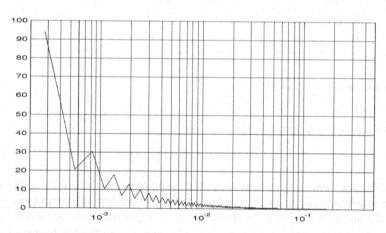

Fig. 5.6 Spectrum of $\hat{u}_r(t)$

5.3 Filter-based ILC

A filter-based ILC scheme is applied with convergence analysis in the frequency domain.

5.3.1 The Schematic of Filter-based ILC

Because of the effectiveness for the performance improvement of the repeated track-ing tasks, ILC has drawn increased attention and many schemes have been devel-oped. In this chapter, a simple ILC scheme will be augmented on top of the exist-ing PI controller with minimal implementation requirements. The block diagram is shown in Fig. 5.7, where y_r is the desired output profile, e_i is the error at the ith iteration, $u_{i,b}$ is the feedback signal at the ith iteration, $u_{i,l}$ is the feedforward signal at the ith iteration, the u_i is the total control signal at the ith iteration, and y_i is the plant output at the ith iteration.

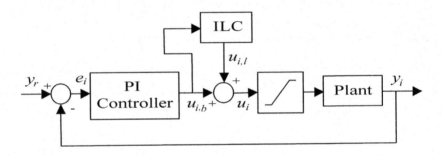

Fig. 5.7 Block diagram of the ILC scheme

The ILC learning update law is given as

$$u_{i+1,l}(k) = u_{i,l}(k) + f(u_{i,b}(k)), \tag{5.3}$$

where k represents the kth time sample of respective signals and $f(\cdot)$ represents some simple numeric operations. It can be seen that the learning control law is es-sentially an integrator along the iteration axis i. In practice, noise is present in the feedback signal and undesirable noise may accumulate in the feedforward signal due to the integration in the iteration domain. For robustness against random pertur-bations, the feedback signal is filtered before applying the learning update law. Note that since $u_{i,b}$ is obtained from the previous operation cycle, it is possible to design $f(\cdot)$ as a zero-phase filter.

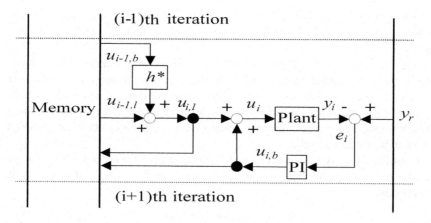

Fig. 5.8 Block diagram of the filter-based ILC scheme

From Fig. 5.8 the learning update law and the overall control input are

$$u_{i+1,l}(k) = u_{i,l}(k) + \gamma h * u_{i,b}(k) \tag{5.4}$$

$$u_i(k) = u_{i,l}(k) + u_{i,b}(k), \tag{5.5}$$

where γ is the filter gain, $h*$ is the filter operator-moving average ($*$ denotes convolution). The non-causal zero-phase filter $\gamma h*$ is a simple moving average with 2 parameters M and γ, related to the filter length and filter gain, respectively

$$\gamma h * u_{i,b}(k) = \frac{\gamma}{2M+1} \sum_{j=-M}^{M} u_{i,b}(k+j). \tag{5.6}$$

Basically, the filter-based ILC attempts to store the desired control signals in the memory bank as the feed-forward signals. With the error convergence, the feedforward signals will tend to the desired control signals so that, in time, it will relieve the burden of the feedback controller. As for all ILC schemes, it is important that the plant output converges to the desired profile along with iterations. This is shown in the convergence analysis.

5.3.2 Frequency-domain Convergence Analysis of Filter-based ILC

Since there exists physical frequency limitation in the control input (PWR), it is sufficient and convenient to provide the frequency-domain convergence analysis in a finite frequency range for the systematic design [13]. Comparing with the time-domain analysis, frequency-domain analysis offers more insight into the effects of the plant, the PI controller and the learning filter on ILC performance, as well as providing the systematic design for M and γ based on the linearized FOPDT plant model.

Since ILC can only be implemented digitally, the frequency analysis should actually be done on sampled-date systems. When the sampling frequency is much faster compared to the system bandwidth, the aliasing problem and the zero-order hold effect are negligible, hence the output signals can be reconstructed as continuous-time signals for the analog plant. Assuming this is so, the analysis can proceed as though for continuous-time systems.

Consider linear systems with $C(j\omega)$ and $P(j\omega)$ as the transfer functions of the feedback controller and the open-loop plant, respectively. The closed-loop transfer function is

$$G(j\omega) = \frac{P(j\omega)C(j\omega)}{1 + P(j\omega)C(j\omega)}. \tag{5.7}$$

Writing the learning updating law (5.4) in the frequency domain gives

$$U_{i+1,l}(j\omega) = U_{i,l}(j\omega) + \gamma H(j\omega)U_{i,b}(j\omega). \tag{5.8}$$

In the following we show that the learning signal $U_{i+1,l}(j\omega)$ will approach the desired control input $u_r(t)$ as the iteration evolves. To show this we need to eliminate $U_{i,b}(j\omega)$. Let us omit $j\omega$ for brevity. From (5.7), we have the following factor

$$
\begin{aligned}
Y_r &= PU_r \\
&= \frac{G(1+PC)}{C}U_r \\
\Rightarrow U_r &= \frac{C}{G(1+PC)}Y_r,
\end{aligned}
\tag{5.9}
$$

where Y_r and U_r are the Fourier transforms of the desired profile $y_r(t)$ and the desired control input $u_r(t)$, respectively. Note that

$$
\begin{aligned}
U_{i,b} &= C(Y_r - Y_i) \\
&= PC(U_r - U_{i,l} - U_{i,b}) \\
\Rightarrow U_{i,b} &= \frac{PC}{(1+PC)}(U_r - U_{i,l}) \\
&= G(U_r - U_{i,l}).
\end{aligned}
\tag{5.10}
$$

Substituting (5.9) into the above relation, $U_{i,b}$ can be written as

$$U_{i,b} = \frac{C}{1+PC}Y_r - GU_{i,l}. \tag{5.11}$$

Thus, (5.8) becomes the following difference equation in the frequency domain consisting of $U_{i,l}$ and U_r

$$U_{i+1,l} = (1 - \gamma HG)U_{i,l} + \gamma HGU_r, \tag{5.12}$$

where H is the Fourier transform of the zero-phase filter $h*$ in (5.4). Iterating the relationship (5.12) by i times, we have

$$\begin{aligned} U_{i+1,l} &= (1 - \gamma HG)^i U_{l,0} + \gamma HGU_r[1 + (1 - \gamma HG) + \cdots + (1 - \gamma HG)^i] \\ &= (1 - \gamma HG)^i U_{l,0} + \gamma HGU_r \frac{1 - (1 - \gamma HG)^i}{1 - (1 - \gamma HG)} \\ &= (1 - \gamma HG)^i U_{l,0} + [1 - (1 - \gamma HG)^i]U_r. \end{aligned} \tag{5.13}$$

Clearly, the convergence condition is

$$|1 - \gamma H(j\omega)G(j\omega)| < 1, \quad \forall \omega \le \omega_b, \tag{5.14}$$

where ω_b is the upper bound of $U_r(\omega)$. When the convergence condition (5.14) is satisfied, the converged value is

$$\lim_{i \to \infty} U_{i,l} = U_r. \tag{5.15}$$

From (5.12), it is interesting to observe that G acts like a filter on U_r, while γH is similar to an "equalizer" used to compensate the "channel" filter. In particular, when H or $G = 0$, the feedforward signal maintains at its initial value, *i.e.* learning will not take place at these filtered frequencies. This is a desired property in higher-frequency bands (above ω_b) dominated by noise. Usually, G is the existing closed-loop control system designed without considering the learning function. Hence, H is the only anti-noise filter, which should remove any frequencies above ω_b, and retain the useful frequency components in U_r. Since $u_r(t)$ is unknown, the frequency-domain design can be based on the information of $\hat{u}_r(t)$, as we shall show in the next section.

In linear systems, if the Nyquist curve of $1 - \gamma HG$ falls within the unit circle for all frequencies in $\hat{U}_r(s)$, then (5.14) is satisfied. Ideally, the Nyquist curve should be close to the origin to give a faster convergence rate of learning for all relevant frequencies.

5.4 Temperature Control of the Water-heating Plant

Experiments are conducted to demonstrate the control performance.

5.4.1 Experimental Setup

Figure 5.9 shows the hardware block diagram of the water-heating plant – PCT23 Process Plant Trainer. In the experiment, the console is interfaced to a PC via the MetraByte DAS-16 AD/DA card. The plant is controlled by a user-written C DOS program in the PC. This program is interrupt driven and serves to command the required control signal to the plant as well as collect and store plant readings. The PI control and reading are done at a rate of 1 Hz, which is more than adequate for the system and the ILC bandwidths. Note that the bandwidth of the desired control input has been calculated based on the estimated $\hat{u}_r$, which is $\omega_b = 0.003\ Hz$ (Fig. 5.6).

5.4.2 Design of ILC Parameters M and γ

Filter-based ILC is used to augment the PI feedback controller. M is designed with two opposing considerations in mind – high-frequency noise rejection and learning rate for $\omega \leq \omega_b$. Since M is the length of the averaging filter, the larger the M, the lower the filter bandwidth, hence the more effective is the noise rejection. On the other hand, as seen in the convergence analysis, H and G, both being low-pass filters, limit the learning rate at higher frequencies. Thus, a large M and hence a small filter bandwidth is detrimental to higher-frequency learning.

For the plant, the bandwidth of G is 0.0045 rad/s, which is slightly above ω_b. To reduce the impact of H on higher frequency learning nearby ω_b, H can be designed such that its bandwidth is slightly larger than G. Setting $M = 10$ and 100 gives filter bandwidths of 0.14 and 0.014 rad/s, respectively. Obviously, $M = 100$ will provide better noise rejection. At the same time, $M = 100$ also gives a bandwidth that is

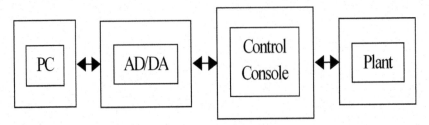

Fig. 5.9 Hardware block diagram of the water-heating plant

slightly larger than G. Thus, $M = 100$ is chosen. The noise-rejection effectiveness of this filter will be verified from empirical results later.

From the bandwidths of G and H, it is seen that the sampling frequency of 1 Hz is adequate to prevent aliasing in the signals. Thus, the frequency-convergence analysis presented in Sect. 5.3 is applicable and the design of γ can be done using Nyquist plots. Using the FOPDT model (5.2) and the PI controller, the Nyquist plot (Fig. 5.10) of $1 - \gamma HG$ with $M = 100$ is obtained for $\gamma = 0.25, 0.50, 0.75$ and 1.00. Note that the size of the curves, *i.e.* the heart-shaped lobe, increases with γ. Also, each curve starts at $1 + 0j$ for $\omega = -\pi f_s$ and traces clockwise back to $1 + 0j$ for $\omega = \pi f_s$, where $f_s = 1$ Hz is the sampling interval.

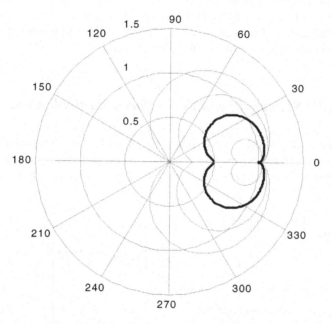

Fig. 5.10 Nyquist plot of $1 - \gamma HG_c$ ($M = 100$)

It can be seen that all the Nyquist curves have portions falling outside the unit circle. Thus, convergence is not guaranteed for $\omega/2\pi > 0.0046, 0.0040, 0.0034$ and $0.0028\,Hz$ for $\gamma = 0.25, 0.5, 0.75$ and 1, respectively. However, Fig. 5.6 shows that the frequency content of the estimated control signal $\hat{U}_r(s)$ is almost negligible for $\omega/2\pi > 0.003$ Hz with $T = 3600$ s. As a tradeoff between stability (location near to the unit circle covering more frequencies) and learning rate (proximity to the origin), $\gamma = 0.5$ is chosen.

5.4.3 Filter-based ILC Results for $\gamma = 0.5$ and $M = 100$

Comparing with the pure PI control (Fig. 5.4), after simple enhancement with the easily implemented ILC, the repetitive system shows vast tracking improvements after 8 iterations (Fig. 5.11). This improvement is also obvious from the RMS error trend shown in Fig. 5.12.

In Fig. 5.11, only a little overshoot and undershoot are seen at the sharp turn in the output profile. With the limited bandwidth of G, the high-frequency components at the sharp turn in $\hat{u}_r(t)$ (Fig. 5.5) are filtered out from $u_{i,l}(t)$. The lack of these components gives the "ringing" at the sharp turn seen in Fig. 5.13, leading to overshoot and undershoot in the output.

From Fig. 5.14, it is also seen that the feedorward signal is relatively smooth and noiseless. This implies that the ILC filter is effective in rejecting noise, making the scheme robust to random perturbations in the system.

5.4.4 Profile Segmentation with Feedforward Initialization

To improve tracking at the sharp turn, a variation on ILC is attempted. From Fig. 5.5, the desired control is piecewise-smooth – consisting of a ramp segment and a level segment. Near the turn, due to the "window averaging" effect of the ILC

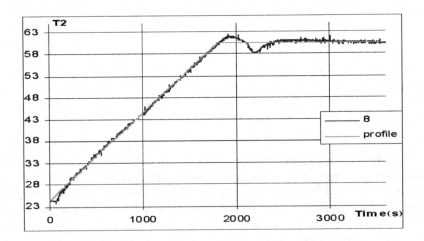

Fig. 5.11 T_2 after 8 iterations ($\gamma = 0.5$ and $M = 100$)

filter, the feedforward signals are derived from these two radically different control efforts – one "correct", the other "wrong". For example, when the averaging window is mainly on the ramp segment but moving towards the turn point, signals taken from ramp segment are the "correct" ones and those from the level segment are the "wrong" ones, and vice versa.

Thus, the original profile is divided into two entirely smooth profiles. ILC will proceed separately for each profile, meaning that the ILC filter will be applied on only one side around the sharp turn. At the same time, the integral part of the PI controller is reset to 0 at the start of each segment. Effectively, it is as though a brand new batch process is started at the turn.

In addition, it is easy to estimate, from the first iteration, the steady-state control effort required for the level segment. At the second iteration, the feedforward signal for the level segment is initialized to this estimate. From then on, ILC proceeds normally for all subsequent iterations, further alleviating any inaccuracy in the estimate.

Compared with Fig. 5.12, the RMS error trend in Fig. 5.15 shows that the improvement from the first to the second iteration is more significant owing to the feedforward initialization. In addition, the error settles to a smaller value.

The smaller error can be attributed to better tracking at the sharp turn as shown in Fig. 5.16. Comparing to Fig. 5.11, there is very much less overshoot and undershoot.

Through segmentation, the filter's "window averaging" effect at the turn is eliminated. In general, the system closed-loop bandwidth still limits the frequency components to be successfully incorporated into the feedforward signal. This limitation is somewhat compensated by initializing the feedforward signal with the control estimate.

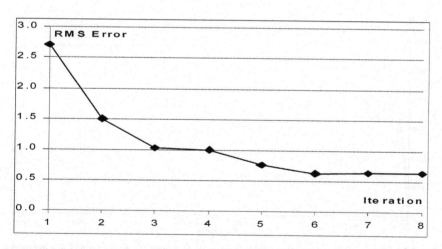

Fig. 5.12 RMS error with the filtered ILC with $\gamma = 0.5$ and $M = 100$

5.4.5 Initial Re-setting Condition

As stated in Chap. 2, a very important property of ILC is that initial plant re-set
is required – the feedforward signal is meaningful only if the plant starts from the
same initial condition in all iterations, this being known as the identical initialization

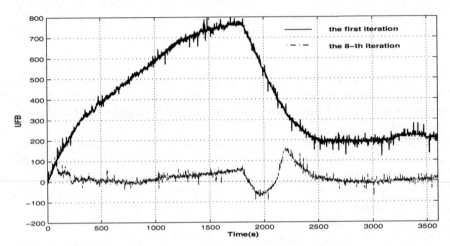

Fig. 5.13 Feedback control signals $u_{i,b}$ at the 1st and 8th iterations with $\gamma = 0.5$ and $M = 100$.
The feedback control signals at the 8th iteration are greatly reduced comparing with the 1st iteration. This is because the tracking error is reduced significantly and the feedforward compensation
dominates the control process after 8 iterations of learning

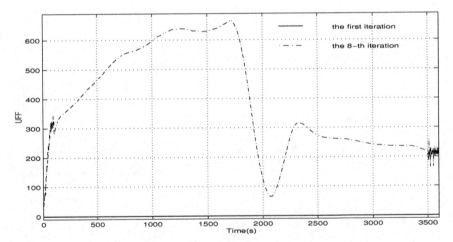

Fig. 5.14 The feedforward learning-based compensation signals $u_{i,l}$ at the 1st and 8th iterations
with $\gamma = 0.5$ and $M = 100$. Comparing with the feedback control signals, the feedforward control
becomes predominant after 8 iterations of learning

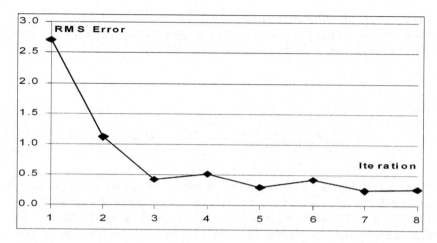

Fig. 5.15 RMS error with the profile segmentation and feed-forward initialization

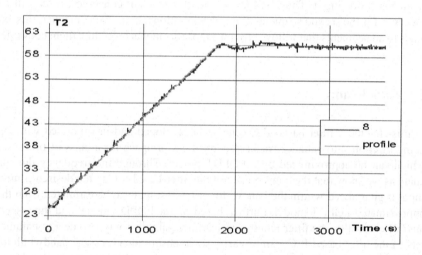

Fig. 5.16 The temperature T_2 profile after 8 iterations with the profile segmentation and feedforward initialization)

condition. Obviously, this is not guaranteed with profile segmentation, except for the first segment.

With convergence given in the first segment, this problem occurs for only the first few iterations, after which the first segment converges to the desired profile so that re-set will be satisfied for the subsequent segment. This is because the endpoint of the 1st segment is exactly the initial point of the 2nd segment. In the same manner, this applies successively to all subsequent segments, if the piecewise-smooth trajectory consists of more than two segments.

Due to the lack of initial reset for the 1st few iterations, a deviation ε will be erroneously incorporated into the feedforward signal during the 1st few iterations. When the initial reset is finally satisfied at the ith iteration, assume that

$$U_{i,l} = U_r + \varepsilon. \qquad (5.16)$$

From (5.12),

$$U_{i+1,l} = (1 - \gamma HG)(U_r + \varepsilon) + \gamma HG U_r$$
$$U_{i+1,l} = U_r + (1 - \gamma HG)\varepsilon. \qquad (5.17)$$

When $|1 - \gamma HG_c| < 1$ for all frequencies of interest, comparing with ε in $U_{i,l}$ it is seen that the deviation in the form of ε is reduced by a factor of $|1 - \gamma HG_c|$ in $U_{i+1,l}$ after 1 iteration. Therefore, the deviation ε will reduce along with further iterations till it finally becomes negligible. Thus, there is no strict requirement for an initial re-set in the first few iterations.

Obviously, the learning convergence in each segment is independent of those segments following it. Thus, as long as the first segment converges, reset will be observed for the second segment after a few iterations. In the same manner, this extends to all segments that follow, provided those segments preceding them converge.

5.5 Conclusion

In this chapter, a filter-based ILC scheme is developed and incorporated with existing PI control to enhance tracking-control performance for a class of processes which can be approximated by a FOPDT model. Through the frequency domain analysis we show that the convergence of the filter-based ILC to the desired control input is guaranteed within the bandwidth of interest that can be estimated from the approximated FOPDT model. Further, based on the FOPDT model and the associated PI controller, the filter length and learning gain, the only two design parameters of the filter-based ILC, can be easily set to ensure the necessary bandwidth for tracking purposes, reject the measurement noise, and achieve a reasonable learning convergence speed. A water-heating system is employed as an illustrative example. A profile-segmentation technique is also employed to improve the tracking performance for piecewise-smooth trajectories. Experimental results clearly show the effectiveness of the proposed method.

5.6 Appendix: The Physical Model of the Water-heating Plant

The plant can be represented in a simplified form as shown Fig.Plant-model [112]. It is assumed that there is no heat loss. In the reservoir tank, the following PDE

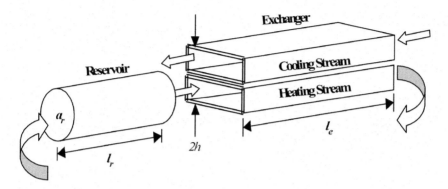

Fig. 5.17 Simplified model of water-heating plant

represents the heating process.

$$\rho_\omega c_\omega \left(\frac{\partial T2}{\partial t} + v_r \frac{\partial T2}{\partial x} \right) = \frac{\text{PWR}(t)}{l_r a_r}, \tag{5.18}$$

where ρ_ω is the density of water, c_ω is the specific heat capacity of water, v_r is the tangential flow velocity in the reservoir, $T2$ is the temperature of reservoir water, and PWR is the power input of the heater.

The heat exchanger is described by the following equations. In this case, the plate exchanger in the physical system is modelled as a counter-flow plate exchanger. For the heating stream and cooling streams respectively, we have

$$\rho_\omega c_\omega \left(\frac{\partial T_h}{\partial t} + v_h \frac{\partial T_h}{\partial x_h} \right) = \frac{c}{h}(T_c - T_h)$$

$$\rho_\omega c_\omega \left(\frac{\partial T_c}{\partial t} + v_c \frac{\partial T_c}{\partial x_c} \right) = \frac{c}{h}(T_c - T_h), \tag{5.19}$$

where T_h and T_c are the heating and cooling stream temperature, respectively, v_h and v_c are the heating and cooling stream flow velocity, respectively, x_h and x_c are the distance from heating and cooling stream inlet, respectively, h is the height of stream, and c is the overall heat-transfer coefficient.

The following initial and boundary conditions apply to the system

Reservoir	Heat exchanger
$T2(x, t = 0) = T_{rm}$	$T_h(x_h, t = 0) = T_c(x_c, t = 0) = T_{rm}$
$T2(x = 0, t) = T_e$	$T_h(x_h = 0, t) = T_r$
	$T_c(x_c = 0, t) = T_{rm}$

where T_{rm} is the room temperature, T_e is the temperature at outlet of heating stream, and T_r is the temperature at the outlet of the reservoir.

Chapter 6
ILC with Robust Smith Compensator: Application to a Furnace Reactor

Abstract The Smith predictor has been used to improve the closed-loop performance for systems with time delays. This chapter proposes a frequency-domain method to design an iterative learning control to further improve the performance of the Smith predictor controller. For a time-invariant plant with multiplicative perturbations and a Smith predictor controller, we derive a sufficient and necessary condition (which has the same form as that of a general robust performance design problem) for the iterative process to converge for all admissible plant uncertainties. In addition, the iterative learning controller under plant uncertainty is designed. An illustrative example demonstrating the main result is presented.[1]

6.1 Introduction

While continuous processing has always been considered as the ideal operation method of a chemical plant, there are a lot of low-volume and high-cost materials obtained in batch-wise form in many chemical and petroleum plants. Improved performance of batch processes is becoming necessary because of competitive markets. Though batch-unit optimal control methods have been developed [103, 119], they are rarely part of everyday industrial practice because of imperfect modelling, unmeasured variables and time delays. As a result of these issues, the degree of automation of many batch units is still very low.

The concept of iterative learning control has been presented by many researchers. ILC provides a method to increase control efficacy by taking advantage of the accumulated batch-to-batch data. Many ILC algorithms have been proposed [4, 14, 21, 67, 71, 100, 143, 147, 148]. The usual approach of the existing studies is to presume a specific learning control scheme in the time domain and then to find the required convergence conditions, and most of the studies focused on only finding open-loop

---

[1] With Elsevier permission to re-use "Iterative learning control with Smith time delay compensator for batch processes," coauthored by Xu, J.-X., Hu, Q.P., Lee, T.H., Yamamoto, S., Journal of Process Control, Vol. 11, no. 3, pp.321–328, 2001.

control signals. In practice, where unexpected disturbances are unavoidable, these algorithms may fail to work.

Most iterative learning control schemes are designed to find purely feedforward action depending wholly on the previous control performance of an identical task. Although the purely feedforward control scheme is theoretically acceptable, it is difficult to apply to real systems without a feedback control due to several reasons. One of the reasons is that it is not robust against disturbances that are not repeatable with respect to iterations. Another reason is that the tracking error may possibly grow quite large in the early stage of learning, though it eventually converges after a number of trials. In addition, an iterative learning control is designed and analyzed with a mathematical model of the plant. Since modelling errors are unavoidable, the real iterative learning control system may violate its convergence condition although the iterative learning control satisfies the condition for nominal plant model. Thus, in real practice, a feedback control is commonly employed along with the iterative learning control for system robustness enhancement and better performance [58, 82]. In these control schemes, the feedback controller ensures closed-loop stability and suppresses exogenous disturbances and the iterative learning controller provides improved tracking performance over a specific trajectory utilizing past control results.

Very few results up to now on the ILC are for dynamics systems with time delay. A feedback-assisted ILC [70] is proposed for chemical batch processes, but this method could be sensitive to process order and time delays, and the general stability analysis is not available. In [42] the possibility of divergence of an ILC is investigated for a plant with time delay. In [96] an ILC algorithm is designed for a class of linear dynamic systems with time delay. However, only uncertainty in time delay is considered in the existing literature.

In the present study, learning control algorithms together with a Smith predictor for batch processes with time delays are proposed and analyzed in the frequency domain. The dynamics of the process are represented by transfer function plus deadtime. Perfect tracking can be obtained under certain conditions. Convergence conditions of the proposed methods are stated and proven. By using the past batch control information, the proposed ILC strategies are able to gradually improve performance of the control system. These results are evaluated through simulation as well as experiments.

6.2 System Description

In this work, a SISO batch process is described by a transfer function $P(s)$ that is assumed to consist of a rational stable transfer function $P_0(s)$ and a dead-time τ. Thus, $P(s)$ is given by

$$P(s) = P_0 e^{-\tau s},$$

and a model of $P(s)$ is described by

$$\hat{P}(s) = \hat{P}_0(s)e^{-\hat{\tau}s},$$

where $\hat{P}_0(s)$ is a model of P_0, and $\hat{\tau}$ is an estimate of τ, the dead-time.

A common configuration of the control system with a Smith predictor is shown in Fig. 6.1, where r is the reference trajectory and $C(s)$ represents the transfer function of the primary controller, which is usually taken to be a conventional PI or PID controller or any lead-lag network. It is well known that $C(s)$ permits higher gains to be used for the structure in Fig. 6.1. This increase in the permitted gains comes from the elimination of the dead-time from the characteristic equation of the closed loop. When a perfect model is available, the overall transfer function, $G(s)$, from r to y is

$$G = \frac{C\hat{P}_0}{1+C\hat{P}_0},$$

which is known as the complementary sensitivity function. Correspondingly, the sensitivity function is

$$S = \frac{1}{1+C\hat{P}_0}.$$

It can be seen that the characteristic equation does not contain dead-time. Stability properties of the system for continuous systems in the presence of modelling errors have been studied [99]. In this study, ILC strategies incorporating this structure will be analyzed and applied to batch processes.

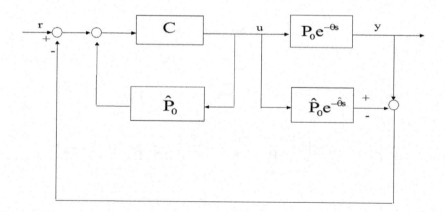

Fig. 6.1 A common configuration with time-delay compensator

Perfect tracking of a continuous process is generally impossible in practice because of modelling uncertainties, unmeasured disturbances and the unreality of compensators. The repeating operations of batch processes make it possible to reduce the tracking error gradually as the batch number increases. The objective of learning control is to progressively achieve perfect tracking that may be written mathematically as

$$\lim_{i \to \infty} \| r(t) - y_i(t) \| = 0 \tag{6.1}$$

for an appropriate norm, where t is defined over the time interval $[0, T]$.

If the process has a time delay τ, the associated time domain has to be shifted to $[\tau, \tau + T]$, and (6.1) becomes

$$\lim_{i \to \infty} \| r(t - \tau) - y_i(t) \| = 0$$

which is the best result we can obtain with feedback control.

6.3 ILC Algorithms with Smith Time-delay Compensator

In this section, an ILC strategy for the case where both the transfer function and the time delay are unknown is considered. Figure 6.2 is a block diagram of the ILC designed in this case and referred to as ILC-1, which is comprised of an iterative learning control law and a Smith predictor.

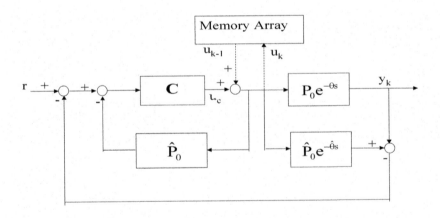

Fig. 6.2 Block diagram of ILC-1

The updating law of the control signal is

$$U_{i+1} = U_i + C[R - (P_0 e^{-\tau s} + \hat{P}_0 - \hat{P}_0 e^{-\hat{\tau}s} U_{i+1}]. \tag{6.2}$$

Multiplying by P on both sides of (6.2) gives the following equation

$$Y_{i+1} = \frac{1}{1 + C(P_0 e^{-\tau s} + \hat{P}_0 - \hat{P}_0 e^{-\hat{\tau}s})} Y_i$$
$$+ \frac{C P_0 e^{-\tau s}}{1 + C(P_0 e^{-\tau s} + \hat{P}_0 - \hat{P}_0 e^{-\hat{\tau}s})} R. \tag{6.3}$$

If we define

$$E_{i+1}^1 = \frac{P_0 e^{-\tau s}}{P_0 e^{-\tau s} + \hat{P}_0 - \hat{P}_0 e^{-\hat{\tau}s}} R - Y_{i+1}, \tag{6.4}$$

then (6.3) can be rewritten as a recursion equation with respect to the tracking error

$$E_{i+1}^1 = Q_1(s) E_i^1, \tag{6.5}$$

where

$$Q_1(s) = \frac{1}{P_0 e^{-\tau s} + \hat{P}_0 - \hat{P}_0 e^{-\hat{\tau}s}} = \frac{S}{1 + ITe^{-\hat{\tau}s}}, \tag{6.6}$$

in which

$$I(s) = \frac{P_0}{\hat{P}_0} e^{\delta \tau s} - 1 \tag{6.7}$$

and

$$\delta \tau = \tau - \hat{\tau}. \tag{6.8}$$

The function $I(s)$ contains all the plant uncertainties and is called the ignorance function [99]. In the ideal case $I(s)$ is identically 0.

Current error E_i can be found from the initial error E_0 by repeated substitution of (6.5)

$$E_i^1 = Q_1(s)^i E_0^1.$$

Clearly, the convergence condition should therefore be

$$|Q_1(s)| < 1, \quad \forall \omega \in [0, \infty).$$

Now we introduce the $\| \cdot \|_2$ norm defined by

$$\|E\|_2^2 \triangleq \frac{1}{2\pi} \int_{-\infty}^{\infty} |E(j\omega)|^2 d\omega = \int_0^{\infty} |e(t)|^2 dt.$$

Taking the norm on both sides of (6.5) gives the following inequality

$$\|E_{i+1}^1\|_2 = \|Q_1(s)E_i^1\|_2 \leq \|Q_1(s)\|_\infty\|E_i^1\|_2, \tag{6.9}$$

where the ∞-norm is defined as

$$\|Q(s)\|_\infty \overset{\triangle}{=} \sup_\omega |Q(j\omega)|.$$

From (6.5) we can see the sequence $\{E_i\}$ converges to 0 if E_0 is bounded and there exists a positive number $M < 1$ such that

$$\|Q_1(s)\|_\infty \leq M < 1.$$

Boundedness of E_0 is guaranteed if the original Smith predictor is stable. In the nominal case where $P = \hat{P}$, (6.9) reduces to

$$\|E_{i+1}^1\|_2 = \|\frac{1}{1+C\hat{P}_0}\|_\infty\|E_i^1\|_2 = \|S\|_\infty\|E_i^1\|_2,$$

which exhibits the main advantages of the primary controller, namely the removal of the time delay from the characteristic equation.

Theoretically, C may be selected regardless of the time delay existing in the system. In practice, however, modelling errors are unavoidable and the controller should accommodate such model–plant mismatches to a certain degree. Although $I(s)$ is not available in practice, it may be constrained to satisfy an upper bound in the frequency domain. To facilitate the analytical derivations, assume

$$|I(j\omega)| \leq \Delta(\omega)$$

for some known $\Delta(\omega)$.

Theorem 6.1. *A sufficient condition for the convergence of the tracking error by* $ILC-1$ *is*

$$\|\,|S| + |\Delta G|\,\|_\infty < 1. \tag{6.10}$$

Proof: From (6.6) we know if

$$\left\|\frac{S}{1+IGe^{-\hat{t}s}}\right\|_\infty < 1, \tag{6.11}$$

the convergence condition is satisfied. Inequality (6.10) implies

$$\|\Delta G\|_\infty < 1$$

and

$$|S| + |\Delta G| < 1$$

at an arbitrary point $j\omega$, therefore

$$\left\| \frac{S}{1 - |\Delta G|} \right\|_\infty < 1. \tag{6.12}$$

Since

$$1 = |1 + IGe^{-\hat{t}s} - IGe^{-\hat{t}s}| \le |1 + IGe^{-\hat{t}s}| + |\Delta G|,$$

we have

$$1 - |\Delta G| \le |1 + IGe^{-\hat{t}s}|.$$

This implies that

$$\left\| \frac{S}{1 - |\Delta G|} \right\|_\infty \le \left\| \frac{S}{1 + IGe^{-\hat{t}s}} \right\|_\infty.$$

This and (6.12) yield (6.11).

When the ILC system is convergent, $\lim_{i \to \infty} E_i^1 = 0$, $i.e.$

$$\lim_{i \to} \left\| \frac{P_0 e^{-\tau s}}{P_0 e^{-\tau s} + \hat{P}_0 - \hat{P}_0 e^{-\hat{t}s}} R - Y_i \right\|_2 = \lim_{i \to \infty} \left\| \frac{I(s) + 1}{I(s) + e^{\hat{t}s}} R - Y_i \right\|_2 = 0.$$

It is clear that Y is approximately equal to R at low frequencies regardless of the presence of modelling uncertainty. In practice, any reference trajectory to be tracked remains in the low-frequency region, and accurate tracking is normally only needed up to the bandwidth of interest. This ILC algorithm does not need an accurate model of the process and can be easily implemented, but perfect tracking may not be obtained. With prior knowledge of the process, other ILC algorithms are proposed to provide perfect tracking in the next section.

6.4 ILC with Prior Knowledge of the Process

Two scenarios with different prior process knowledge will be explored.

6.4.1 ILC with Accurate Transfer Function ($P_0 = \hat{P}_0$)

In the case where P_0 can be accurately estimated, $i.e.$ $P_0 = \hat{P}_0$, another ILC strategy is proposed in this work such that perfect tracking can be obtained. Figure 6.3 shows the block diagram of the ILC algorithm (ILC-2). From this structure, we get

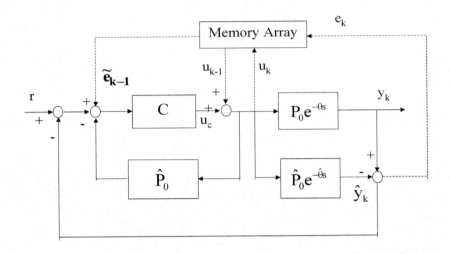

Fig. 6.3 Block diagram of ILC-2 ($P_0 = \hat{P}_0$)

$$U_{i+1} = U_i + C[R - (P_0 e^{-\tau s} + \hat{P}_0 - \hat{P}_0 e^{-\hat{t}s})U_{i+1} + \tilde{E}_i, \tag{6.13}$$

where

$$\tilde{E}_i = Y_i - \hat{Y}_i = (P_0 e^{-\tau s} - \hat{P}_0 e^{-\hat{t}s})U_i. \tag{6.14}$$

Substituting (6.14) into (6.13) gives

$$U_{i+1} = \frac{1 + C(P_0 e^{-\tau s} - \hat{P}_0 e^{-\hat{t}s})}{1 + C(P_0 e^{-\tau s} + \hat{P}_0 - \hat{P}_0 e^{-\hat{t}s})} U_i$$
$$+ \frac{C}{1 + C(P_0 e^{-\tau s} + \hat{P}_0 - \hat{P}_0 e^{-\hat{t}s})} R. \tag{6.15}$$

Multiplying by $P_0 e^{-\tau s}$ on both sides of (6.15) gives the following recursion equation

$$Y_{i+1} = \frac{1 + C(P_0 e^{-\tau s} - \hat{P}_0 e^{-\hat{t}s})}{1 + C(P_0 e^{-\tau s} + \hat{P}_0 - \hat{P}_0 e^{-\hat{t}s})} Y_i$$
$$+ \frac{C P_0 e^{-\tau s}}{1 + C(P_0 e^{-thetas} + \hat{P}_0 - \hat{P}_0 e^{-\hat{t}s})} R. \tag{6.16}$$

If we define

$$E_{i+1}^2 \triangleq R e^{-\tau s} - Y_{i+1}, \tag{6.17}$$

and since $P_0 = \hat{P}_0$, (6.16) can be rewritten as a recursion equation with respect to the control error

$$E_{i+1}^2 = Q_2(s)E_i^2$$

where

$$Q_2(s) = \frac{1 + C(\hat{P}_0 e^{-\tau s} - \hat{P}_0 e^{-\hat{\tau}s})}{1 + C(\hat{P}_0 e^{-\tau s} - \hat{P}_0 e^{-\hat{\tau}s}) + C\hat{P}_0} = \frac{S + IGe^{-\hat{\tau}s}}{1 + IGe^{-\hat{\tau}s}}.$$

Therefore, the convergence condition of (6.18) is satisfied if

$$\left\| \frac{S + IGe^{-\hat{\tau}s}}{1 + IGe^{-\hat{\tau}s}} \right\|_\infty < 1. \tag{6.18}$$

Theorem 6.2. *A sufficient condition for the convergence of the tracking error of (6.18) is*

$$\| |S| + 2|\Delta G| \|_\infty < 1. \tag{6.19}$$

Proof: Inequality (6.19) implies

$$\|\Delta G\|_\infty < 1 \quad \text{and} \quad \left\| \frac{|S| + |\Delta G|}{1 - |\Delta G|} \right\|_\infty < 1. \tag{6.20}$$

Since

$$|S| + |\Delta G| \geq |S| + |IGe^{-\hat{\tau}s}| \geq |S + IGe^{-\hat{\tau}s}|,$$

and

$$1 - |\Delta G| \leq 1 - |IG| \leq |1 + IGe^{-\hat{\tau}s}|,$$

we have

$$\left\| \frac{S + IGe^{-\hat{\tau}s}}{1 + IGe^{-\hat{\tau}s}} \right\| \leq \left\| \frac{|S| + |\Delta G|}{1 - |\Delta G|} \right\|_\infty. \tag{6.21}$$

Equation (6.18) can be obtained from (6.20) and (6.21).

From this theorem, $\lim_{i \to \infty} \|E_i^2\|_2 = 0$ if (6.19) is satisfied. Therefore, in this case, perfect tracking is obtained. If P_0 is not accurate, *i.e.*, $P_0 \neq \hat{P}_0$, the tracking error in (6.17) will change to

$$E_i^2 = \frac{P_0}{\hat{P}_0} Re^{-\tau s} - Y_i. \tag{6.22}$$

This means that the tracking performance depends on the accuracy of the transfer function.

6.4.2 ILC with Known Upper Bound of the Time Delay

An ILC algorithm shown in Figure 6.4 (referred to as ILC-3) is also proposed for perfect tracking in this study. In this ILC algorithm, only the bound of the time delay

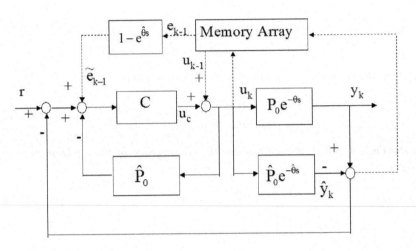

Fig. 6.4 Block diagram of ILC-3 ($\tau = \hat{\tau}$)

is needed. Assume $\tau \in [0, \tau_{\max}]$. We consider the worst case by choosing $\hat{\tau} = \tau_{\max}$. Based on Fig. 6.4, the following equation can be obtained.

$$U_{i+1} = U_i + C[R - (P_0 e^{-\tau s} + \hat{P}_0 - \hat{P}_0 e^{-\hat{\tau}s})U_{i+1} + \tilde{E}_i,$$

where

$$\tilde{E}_i = (Y_i - \hat{Y}_i)(1 - e^{-\tau s}).$$

Since Y_i and $\hat{Y}_i$ are pre-acquired signals, $Y_i e^{-\hat{\tau}s}$ and $\hat{Y}_i e^{-\hat{\tau}s}$ are available. Analogous to the procedures in the preceding section, we can derive

$$Y_{i+1} = \frac{1 + C(P_0 e^{-\tau s} - \hat{P}_0 e^{-\hat{\tau}s} + \hat{P}_0 - P_0 e^{-(\tau - \hat{\tau})})}{1 + C(P_0 e^{-\tau s} - \hat{P}_0 e^{-\hat{\tau}s} + \hat{P}_0)} Y_i + \frac{CP_0 e^{-\tau s}}{1 + C(P_0 e^{-\tau s} - \hat{P}_0 e^{-\hat{\tau}s} + \hat{P}_0)} R,$$

which can be written as the following recursion equation

$$E_{i+1}^3 = Q_3(s)E_i^3,$$

where

$$E_i^3 = Re^{-\hat{\tau}s} - Y_i$$

$$Q_3(s) = \frac{1+C(P_0e^{-\hat{t}s} - \hat{P}_0e^{-\hat{t}s} + \hat{P}_0 - P_0e^{-(\tau-\hat{t})}}{1+C(P_0e^{-\tau s} - \hat{P}_0e^{-\hat{t}s} + \hat{P}_0)}$$
$$= \frac{S+IG(e^{-\hat{t}s} - 1)}{1+IGe^{-\hat{t}s}}.$$

It is clear that the convergence condition of ILC-3 is satisfied iff

$$\left\| \frac{S+IG(e^{-\hat{t}s} - 1)}{1+IGe^{-\hat{t}s}} \right\|_{\infty} < 1. \tag{6.23}$$

Theorem 6.3. *If*

$$\| |S| + 3|\Delta G| \|_{\infty} < 1, \tag{6.24}$$

perfect tracking can be obtained by ILC-3.

Proof: Assume (6.24), or equivalently,

$$\| \Delta G \|_{\infty} < 1 \quad and \quad \left\| \frac{|S|+2|\Delta G|}{1-|\Delta G|} \right\|_{\infty} < 1. \tag{6.25}$$

Obviously,

$$|S+IG(e^{-\hat{t}s} - 1)| \leq |S+IGe^{-\hat{t}s}| + |IG| \leq |S| + 2|\Delta G|, \tag{6.26}$$
$$1-|\Delta G| \leq 1 - |IG| \leq |1+IGe^{-\hat{t}s}|. \tag{6.27}$$

Based on (6.25)–(6.27), (6.23) is easily obtained.

6.5 Illustrative Examples

To show the effectiveness of ILC, we first show simulation results, then experimental tests on a batch reactor which is a temperature control problem of a furnace.

6.5.1 Simulation Studies

In this part, numerical examples are given to demonstrate the behavior of the proposed ILC algorithms. We assumed the following first-order process with time delay is repeated at every batch.

$$P(s) = \frac{1}{s+1}e^{-s}.$$

For the first two cases (ILC-1 and ILC-2), PD controllers

$$C^1(s) = s+1$$
$$C^2(s) = 0.5(s+1)$$

were used separately, while for ILC-3, a PI controller was tested

$$C^3(s) = 1 + \frac{1}{s}.$$

The nominal models for the three cases are, respectively,

$$\hat{p}^1 = \frac{2}{1.5s+1}e^{-1.2s}$$
$$\hat{p}^2 = \frac{1}{s+1}e^{-2s}$$
$$\hat{p}^3 = \frac{10}{8s+7}e^{-1.2s}.$$

All these controllers and models are selected such that the stability conditions are satisfied. After some calculations, we have $\|Q_1\|_\infty = 0.83$, $\|Q_2\|_\infty = 0.78$ and $\|Q_3\|_\infty = 1$. Although $\|Q_3\|_\infty = 1$, it only occurs when the frequency is sufficiently high. Figure 6.5 is the magnitude plot of Q_1, Q_2 and Q_3. Therefore, all the convergence conditions are satisfied. All the ILC control systems are initiated with the PD or PI control result in the first run. In Fig. 6.6, the performance of ILC-1 is shown. Although significant modelling errors exist, ILC-1 gives satisfactory performance. The performance of ILC-2 and ILC-3 is demonstrated in Fig. 6.7 and Fig. 6.8, respectively. As expected, almost perfect tracking is obtained by these two strategies after some iterations of learning.

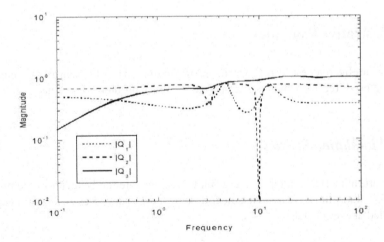

Fig. 6.5 Magnitude plot of $Q(i\omega)$

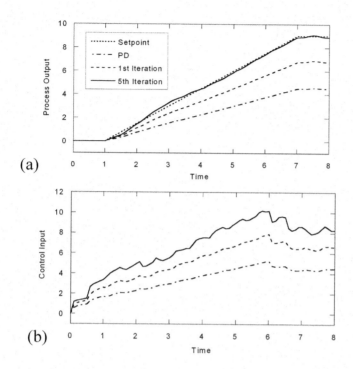

Fig. 6.6 Simulation results of ILC-1: (a) process output and (b) control input

6.5.2 Experiment of Temperature Control on a Batch Reactor

To assess the practical applicability of the proposed ILC-1, experiments were conducted in a batch reactor. Figure 6.9 is a schematic diagram of the experimental setup. The setup consists of a reactor, a heater, an R-type thermocouple, an amplifier, a thyristor unit and an analog-to-digital and digital-to-analog conversion (AD/DA) card, a personal computer and cooling-water system. The control objective was to track the desired temperature curve despite modelling errors.

In the experiment, alumina powder was produced in the reactor by pyrolysis of NH4-alum. Temperature increment speeds of the reactor during the experiment determine the characteristics of produced alumina powder. A model of the setup was obtained using open-loop identification in a Matlab/Simulink environment.

$$\hat{P} = \frac{624}{243s+1}e^{-30s}.$$

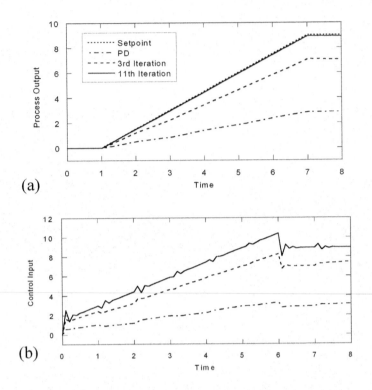

Fig. 6.7 Simulation results of ILC-2: (a) process output and (b) control input

Input delay was intentionally added to make the control task more difficult. The first run of ILC-1 was initiated with the result of PI controller with parameter $K_c = 0.003$ and $T_i = 30$ that were obtained using the relay tuning method.

The experimental results of PI and ILC-1 are shown in Fig. 6.10 and Fig. 6.5.2. A PI controller gives quite good performance owing to well-tuned PI parameters. ILC-1 further improves the performance. The control error was reduced after the learning control started to work, and the control error further reduced after the second batch. After the 3rd iteration of learning, the tracking error is reduced significantly and sufficiently precise temperature control was attained over the whole reaction period.

6.6 Conclusion

The problem of iterative learning control of batch processes with time delays has been addressed. The processes were represented by a transfer function plus dead-time. Convergence conditions were derived in the frequency domain. Perfect tracking can be obtained under certain conditions. Since the proposed ILC schemes are

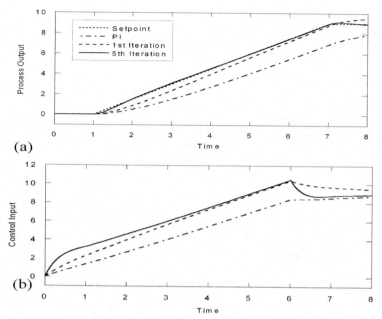

(a)

(b)

Fig. 6.8 Simulation results of ILC-3: (a) process output and (b) control input

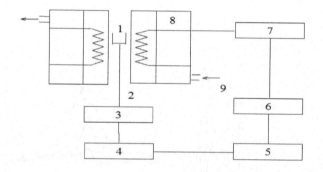

1: Reactor 2: Thermocouple 3: Amplifier 4: A/D converter 5: Computer

6: D/A Converter 7: Thyristor 8: Heater 9: Cooling water

Fig. 6.9 Experimental set-up

directly attached to the existing control systems, they can easily be applied to batch processes to improve control performance. Simulations and experimental results on the temperature control of a reactor show the effectiveness of the proposed methods.

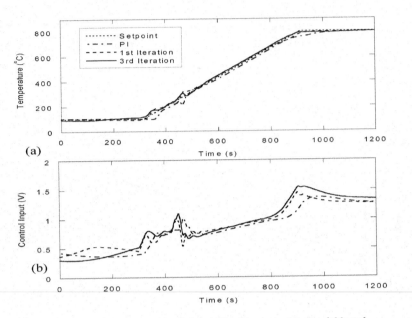

Fig. 6.10 Experimental results: (a): controlled variable, (b): manipulated variable and

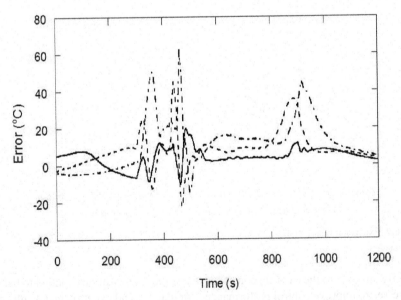

Fig. 6.11 Experimental results: (a) controlled variable, (b) manipulated variable and (c) tracking error

Chapter 7
Plug-in ILC Design for Electrical Drives: Application to a PM Synchronous Motor

Abstract Permanent-magnetic synchronous motor (PMSM) drives are widely used for high-performance industrial servo and speed-regulation applications where torque smoothness is an essential requirement. However, one disadvantage of PMSM is parasitic torque pulsations, which induce speed oscillations that deteriorate the drive performance particularly at low speeds. To suppress these speed ripples, two ILC algorithms implemented in the time domain and frequency domain, respectively, are proposed in this chapter. Although a conventional PI speed controller does suppress speed ripples to a certain extent, it is not adequate for many high-performance applications. Thus, the proposed plug-in ILC controller is applied in conjunction with a PI speed controller to further reduce the periodic speed ripples. Experimental verification of the two algorithms is carried out, and test results obtained demonstrate that the algorithm implemented in the frequency domain has better performance in reducing speed ripples than that implemented in the time domain because of the elimination of the forgetting factor that is indispensable for robustness in the time-domain learning method.[1]

7.1 Introduction

PM synchronous motor drives are widely used in robotics, machine tools and other high-performance industrial applications to servo and speed regulation. PMSMs are preferred over the traditional brush-type DC motors because of the absence of mechanical commutators, which reduces mechanical wear and tear of the brushes and increases the life span of the motor. As compared to induction motors, PMSMs are still favored for high-performance servo applications because of their high efficiency, power density and torque-to-inertia ratio, which make them a suitable choice for variable-speed direct-drive applications.

---

[1] With IEEE permission to re-use "Speed ripples minimization in PM synchronous motors using iterative learning control," coauthored Qian, W.Z., Panda, S.K., Xu, J.-X., IEEE Transactions on Energy Conversion, Vol. 20, no.1, pp.53–61, 2005.

However, the main disadvantage of PMSMs is the parasitic torque pulsations [54]. The presence of these torque pulsations results in instantaneous torque that pulsates periodically with rotor position. These pulsations are reflected as periodic oscillations in the motor speed, especially for low-speed operation. At higher operating speeds, these torque pulsations are naturally filtered off by the rotor and load inertias and therefore, are not reflected back in the motor speed. But in the absence of mechanical gears in direct-drive servos, the motor drive has to operate at low speeds. These speed oscillations severely limit the performance of the servo especially in high-precision tracking applications. Moreover, the oscillations produce undesirable mechanical vibration on the load side.

There are various sources of torque pulsations in a PMSM such as the cogging, flux harmonics, errors in current measurements and phase unbalancing. In view of the increasing popularity of PMSMs in industrial applications, the suppression of pulsating torques has received much attention in recent years. Broadly speaking, these techniques can be divided into two groups: one focusing on the improvement of motor design and the other emphasizing the use of active control of stator current excitations [54]. From the motor-design viewpoint, skewing the stator lamination stacks or rotor magnets, arranging proper winding distribution and other motor design features reduce cogging torque to a certain degree but do not completely eliminate it [121]. Moreover, special machine-design techniques additionally increase the complexity in the production process, which results in higher machine cost.

The second approach, which is our interest, concentrates on using an additional control effort to compensate these periodic torque pulsations. One of the earliest method proposed in [49, 40] is to use pre-programmed stator current excitation to cancel torque harmonic components. However, in such a method, sufficiently accurate information on the PMSM parameters, in particular the characteristics of torque ripples is required, and a small error or variations in parameters can lead to an even higher torque ripple due to the open-loop feedforward control. In view of the inherent limitations of these open-loop control algorithms, alternative approaches applying closed-loop control algorithms with online estimation techniques (*e.g.*, a self-commissioning algorithm [44] and an adaptive control algorithm [98]) to achieve torque-pulsation minimization have been proposed. These real-time control algorithms are implemented either in speed or in current (torque) loops. In torque-control algorithms, one popular way is to regulate torque by using on-line estimated torque based on electrical-subsystem variables (currents and voltages) only. Various algorithms have been proposed for instantaneous torque estimation [78, 80, 26, 24]. On the other hand, this approach can be used only for those torque-ripple components that are observable from the electrical subsystem – ripples due to mechanical part (*e.g.*, cogging torque and load oscillations) cannot reflect in electrical subsystem variables, and hence are uncontrollable [98]. An alternative technique is to use a torque transducer with high bandwidth output to measure the real-time torque signal, which can then be applied as the feedback information. This, however, increases the cost of the drive system. While all of the techniques described above seek to attenuate torque pulsations by adjusting the stator current excitations, an alternative approach relies on a closed-loop speed regulator to accomplish the same objective indirectly, *i.e.* the attenuation of torque ripples [80, 98]. All possible sources of torque ripples are observable from rotor speed, hence this method has potential for complete torque-ripple minimization. However, the quality of speed feedback and the slow dynamics of the outer speed loop limit the dynamic performance of the algorithm.

In this chapter, two iterative learning control algorithms implemented in the time domain as well as in the frequency domain, respectively, have been proposed with the objective of minimizing periodic speed ripples originated by torque pulsations. Regardless of the difference between their control structures, both control algorithms have the same drive configuration. The proposed ILC controller is applied in conjunction with the conventional PI speed controller, which provides the main reference current. During steady state, the proposed controller generates the compensation current that together with the main reference current is utilized to minimize the speed ripples. Conventional PI current controllers are used in the inner control loops to generate the control voltages to shape PWM signals. The performances of both ILC control algorithms have been evaluated through extensive experimental investigations. Test results obtained demonstrate improvements in the steady-state speed response and therefore validate the effectiveness of both ILC algorithms. A comparison between the two algorithms also shows the advantage of the algorithm implemented in the frequency domain, because of the elimination of the forgetting factor used for robustness in the time-domain learning method, which further enhances the effectiveness in suppressing speed ripples.

The chapter is organized as follows. In Sect. 7.2, the mathematical model of the PMSM is given. Section 7.3 briefly describes the sources of torque pulsations and the resultant speed ripples in PMSM drive. In Sect. 4, the proposed ILC algorithms are explained, and PMSM drive setup used in experiments is introduced in Sect. 5. Section 6 presents and discusses the experimental results. Finally, some concluding remarks are made in Sect. 7.

7.2 PMSM Model

With assumptions that the PMSM is unsaturated and eddy currents and hysteresis losses are negligible, the stator d, q-axes voltage equations of the PMSM in the synchronous rotating reference frame are given by

$$\frac{di_{ds}}{dt} = -\frac{R}{L_d}i_{ds} + \frac{L_q}{L_d}\omega_e i_{qs} + \frac{1}{L_d}v_{ds} \tag{7.1}$$

$$\frac{di_{qs}}{dt} = -\frac{R}{L_q}i_{qs} - \frac{L_d}{L_q}\omega_e i_{ds} - \frac{\omega_e}{L_q}\psi_{dm} + \frac{1}{L_q}v_{qs}, \tag{7.2}$$

where i_{ds} and i_{qs} are the d, q-axes stator currents, v_{ds} and v_{qs} are the d, q-axes voltages, L_d and L_q are the d, q-axes inductances, while R and ω_e are the stator resistance and electrical angular velocity, respectively [66]. The flux linkage ψ_{dm} is due to rotor magnets linking the stator. It has been further assumed that as the surface-mounted PMSM is non-salient, L_d and L_q are equal and are taken as L.

Using the method of field-oriented control of the PMSM, the d-axis current is controlled to be zero to maximize the output torque. The motor torque is given by

$$T_m = \frac{3}{2}\frac{p}{2}\psi_{dm}i_{qs} = k_t i_{qs}, \tag{7.3}$$

in which $k_t (= \frac{3}{2}\frac{p}{2}\psi_{dm})$ is the torque constant and p is the number of poles in the motor.

The equation of the motor dynamics is

$$\frac{d\omega_m}{dt} = -\frac{B}{J}\omega_m + \frac{k_t}{J}i_{qs} - \frac{T_l}{J}, \tag{7.4}$$

where ω_m is the mechanical rotor speed, T_l is the load torque, B is the frictional coefficient and J is the total inertia (motor and load).

7.3 Analysis of Torque Pulsations

A. Flux Harmonics

Due to the non-sinusoidal flux-density distribution in the air gap, the resultant flux linkage between the permanent magnet and the stator currents contains harmonics of the order of 5, 7, 11, $\cdots$ in the a-b-c frame (triple harmonics are absent in Y-connected stator windings) [26]. In the synchronous rotating reference frame, the corresponding harmonics appear as the 6th, 12th and the multiples of 6th-order harmonic components, and can be expressed as

$$\psi_{dm} = \psi_{d0} + \psi_{d6}\cos 6\theta_e + \psi_{d12}\cos 12\theta_e + \cdots, \tag{7.5}$$

where ψ_{d0}, ψ_{d6} and ψ_{d12} are the DC, 6th and 12th harmonic terms of the d-axis flux linkage, respectively, while θ_e is the electrical angle. Combining (7.3) and (7.5), we get:

$$\begin{aligned} T_m &= T_0 + T_6\cos 6\theta_e + T_{12}\cos 12\theta_e + \cdots \\ &= T_0 + \triangle T_{m,6} + \triangle T_{m,12} + \cdots, \end{aligned} \tag{7.6}$$

where T_0, T_6 and T_{12} are the DC component, 6th and 12th harmonic torque amplitudes respectively. Equation 7.6 indicates that the 6th and 12th torque harmonics, produced mainly due to the on-sinusoidal flux distribution, are periodic in nature.

B. Current Offset Error

The DC offset in stator current measurements also leads to pulsating torque [25]. Stator currents are measured and transduced into voltage signals by current sensors and then transformed into digital form by A/D converters. The presence of any unbalanced DC supply voltage in the current sensors and inherent offsets in the analog electronic devices give rise to DC offsets. Letting the DC offsets in the measured currents of phases a and b be Δi_{as} and Δi_{bs}, respectively, the "measured" q-axis current i_{qs-AD} can be expressed as:

$$i_{qs-AD} = i_{qs} + \Delta i_{qs}, \tag{7.7}$$

where

$$\Delta i_{qs} = \frac{2}{\sqrt{3}}\cos(\theta_e + \alpha)\sqrt{\Delta i_{as}^2 + \Delta i_{as}\Delta i_{bs} + \Delta i_{bs}^2}, \tag{7.8}$$

and α is a constant angular displacement and dependent on Δi_{as} and Δi_{bs}. As $\theta_e = 2\pi f_s t$, it is shown that Δi_{qs} oscillates at the fundamental electrical frequency. Assuming the measured currents exactly follow the reference currents, the actual motor current is given by

$$i_{qs} = i_{qs-AD} - \Delta i_{qs} = i_{qs}^* - \Delta i_{qs}. \tag{7.9}$$

Using (7.3) and (7.9), we get

$$T_m = k_t(i_{qs}^* - \Delta i_{qs}) = T_m^* - \Delta T_{m,1} \tag{7.10}$$

where $\Delta T_{m,1} = k_t \Delta i_{qs}$ is a torque oscillation at the fundamental frequency f_s. From (7.8) $\Delta T_{m,1}$ can be obtained as

$$\Delta T_{m,1} = k_t \frac{2}{\sqrt{3}}\cos(\theta_e + \alpha)\sqrt{\Delta i_{as}^2 + \Delta i_{as}\Delta i_{bs} + \Delta i_{bs}^2}. \tag{7.11}$$

Equation 7.11 shows that the offsets in current measurement give rise to a torque oscillation at the fundamental frequency f_s.

C. Current Scaling Error

The output of the current sensor must be scaled to match the input of the A/D converter, and in the digital form, the controller re-scales the value of the A/D output to obtain the actual value of the current. As such, scaling errors of the currents are inevitable [25]. Again, assuming ideal current tracking, the measured phase currents are

$$i_{as-AD} = i_{as}^* = I\cos\theta_e \tag{7.12}$$

$$i_{bs-AD} = i_{bs}^* = I\cos\left(\theta_e - \frac{2\pi}{3}\right). \tag{7.13}$$

Denoting the scaling factors of the phases a and b currents as K_a and K_b, respectively, the phase currents as seen from the controller are

$$i_{as} = I\cos\theta_e / K_a \tag{7.14}$$

$$i_{bs} = I\cos\left(\theta_e - \frac{2\pi}{3}\right) / K_b. \tag{7.15}$$

From similar analysis as in the current offset error, $\triangle i_{qs}$ can be evaluated as:

$$\triangle i_{qs} = i_{qs-AD} - i_{qs}$$
$$= \left(\frac{K_a - K_b}{K_a K_b}\right) \frac{I}{\sqrt{3}} \left[\cos\left(2\theta_e + \frac{\pi}{3}\right) + \frac{1}{2}\right]. \qquad (7.16)$$

From (7.10) and (7.16), the torque error is

$$\triangle T_{m,2} = \left(\frac{K_a - K_b}{K_a K_b}\right) \frac{k_t I}{\sqrt{3}} \left[\cos\left(2\theta_e + \frac{\pi}{3}\right) + \frac{1}{2}\right]. \qquad (7.17)$$

Equation 7.17 shows that the scaling error causes the torque to oscillate at twice the fundamental frequency $2f_s$.

These analyses indicate that the electromagnetic torque consists of a DC component together with the 1st, 2nd, 6th and 12th harmonic components. The control objective is to suppress these periodic torque ripples by the proposed ILC algorithms indirectly.

D. Speed Ripples Caused by Torque Pulsations
The plant transfer function between the motor speed and the torque is

$$\omega_m(s) = \frac{T_m(s) - T_l(s)}{Js + B}, \qquad (7.18)$$

where $T_m = f(\psi_{dm}, i_{qs}, \omega_m)$. It can be seen that the speed would oscillate at the same harmonic frequencies as those of $\triangle T_m$, especially at low operating speeds. In order to minimize the speed ripples, the sources of these speed oscillations – torque pulsations need to be minimized. However, for suppression of torque ripples it is necessary to measure or estimate the instantaneous torque, which makes the drive system expensive (by using a torque transducer) or complicated (by using a torque observer). Therefore, in the proposed algorithm as described in the next section, speed information that is already available for the closed-loop speed-control purpose, is utilized to compensate for torque ripples indirectly. Consequently, the suppression of torque ripples leads to reduction in speed ripples.

7.4 ILC Algorithms for PMSM

Iterative learning control is an approach to improving the tracking performance of systems that operate repetitively over a fixed time interval. In this work we explore two ways of implementing ILC in the time domain and the frequency domain.

There are three issues that arise in general for electrical drive systems when ILC is applied.

First, the electrical drive system is running in the repetitive mode continuously instead of the repeated mode that stops and restarts. Therefore, it is a repetitive control task. From cycle to cycle the identical initialization condition only holds for angular displacement θ_e but does not hold for other states such as the speed ω_m, currents i_{qs} and i_{ds}. In principle, repetitive control [41, 76] or repetitive learning control [30, 160] instead of ILC should be used for such applications. However, repetitive control algorithms were developed mainly for linear processes, and repetitive learning control algorithms require some conditions that are not satisfied in PMSM. As is discussed in Sect. 2.3, ILC can be used for repetitive tasks.

Second, we note that the ripples in PMSM are cyclic with respect to the electrical angle θ_e. Thus, this task is different from previous ILC problems where the process repeatability is defined with respect to the time. To address this problem, we can conduct learning according to the electrical angle that is cyclic, and update the present cycle control signals using the previous cycle control and error signals. In other words, this is a spatial ILC algorithm.

Third, the implementation of this spatial ILC in sampled-data systems will encounter a new problem – a fixed sampling mechanism may not provide the control and error signals at the desired angular position that varies when the PMSM speed changes. For instance, the updating is carried out at every 1 degree of the electrical angle, but the sampling instance could fall on a position with a fractional value of a degree. A solution to this problem is to use interpolation between two sampled positions [137]. In this work, we use a sufficiently high sampling frequency of 1250 Hz, and do not use interpolation. This is because the interpolated values do not deviate much from the sampled values when the sampling interval is sufficiently small. Further, when the PMSM works at the steady state, the speed does not change drastically. As such, the learning can still be carried out in the time domain, because the angular displacement is linearly proportional to the time t.

7.4.1 ILC Controller Implemented in Time Domain

In our algorithm, we have adopted a P-type learning controller employing the combined PCL and CCL algorithm. The P-type algorithm is simple to implement, unlike in the D-type since differentiation of the speed signal is unnecessary, hence noise build-up in input update caused by the differentiation of the speed signal can be avoided. In [5], a mathematically rigorous treatment of the robustness and convergence of the P-type learning control algorithm is given. It is found that the introduction of a forgetting factor α increases the robustness of the P-type algorithm against noise, initialization error and fluctuation of system dynamics. The proposed ILC algorithm in the time domain is illustrated in Fig. 7.1, and the following learning law is used:

$$u_{i+1}(t) = (1 - \alpha)u_i(t) + \beta_l e_i(t) + \beta e_{i+1}(t), \tag{7.19}$$

where $i = 1, 2, 3 \cdots$ is the iteration or cycle number, u_i is the control signal, namely, reference compensation q-axis current generated from ILC, e_i is the speed error signal or $\omega_m^* - \omega_m$, α is the forgetting factor, β_l and β are the PCL and CCL gains, respectively.

To determine the learning gain β_l, assuming perfect tracking of the inner-loop current controller:

$$T_m(t) = T_m^*(t) = k_t i_{qs}^*(t), \tag{7.20}$$

and substituting (7.20) into (7.4) yields:

$$\frac{d\omega_m(t)}{dt} = -\frac{B}{J} + \frac{k_t}{J} i_{qs}^*(t) - \frac{1}{J} T_l. \tag{7.21}$$

For the purpose of convergence, the following condition must hold according to [14] or Chap. 2

$$\left\| 1 - \frac{k_t}{J} \beta_l \right\| < 1. \tag{7.22}$$

Denoting $0 < k_t/J \le |k_t/J|_{max}$, the inequality in (7.22) can be solved as

$$0 < \beta_l < \frac{2}{|k_t/J|_{max}}. \tag{7.23}$$

From the knowledge of the range of d-axis flux linkage ψ_{dm} (since k_t depends on ψ_{dm}) and the total inertia J, the learning gain β_l can be determined. To achieve a faster rate of convergence, the value of β_l should approach $2|k_t/J|_{max}^{-1}$. However, a conservative choice of β_l that ensures stability and reasonably fast convergence would suffice. From Chap. 2, the CCL gain, β will improve the convergence in the iteration axis of the learning controller. However, having β that is too large will cause over-amplification of the error or noise signals in the input updating and the corresponding control output will tend to be large, leading to the eventual divergence of the output. As before, a conservative choice of β such that $\beta \le \beta_l$ would suffice to produce good results.

7.4.2 ILC Controller Implemented in Frequency Domain

Unfortunately, the introduction of the forgetting factor in the time-domain learning algorithm only ensures the tracking errors within a certain bound, and further improvement is limited with the consideration of robustness. Under such conditions, we further implement learning control in the frequency domain by means of Fourier-series expansion [149]. Fourier-series-based learning enhances the robustness of iterative learning and in the meanwhile maintains the possibility of reducing tracking errors to zero, theoretically. Note that there always exists system noise or other small non-repeatable factors in the system. Accumulation of those components contained in $u_i(t)$ may degrade the approximation precision of the controller for each new trial. A Fourier-series-based learning mechanism, on the other hand, updates coefficients of the learned frequency components over the entire learning

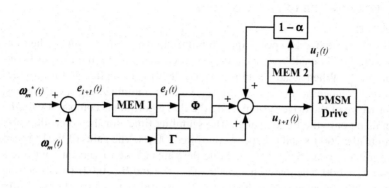

Fig. 7.1 Block diagram of the ILC algorithm implemented in the time domain

period according to (7.25), which takes an averaging operation on noise and is able to remove the majority of noise and non-repeatable factors. Consequently, there is no need to introduce the forgetting factor in this algorithm. The repeatability of speed ripples implies that only countable integer multiples of the frequency are involved. This ensures the feasibility of constructing componentwise learning in the frequency domain. Here is the control law:

$$u_{i+1}(t) = FT(u_i(t)) + \beta_l e_i(t) + \beta e_{i+1}(t) \tag{7.24}$$

$$FT(u_i(t)) = \psi^T \cdot \frac{2}{T} \int_0^T u_i(t)\psi_1 dt, \tag{7.25}$$

where

$$\psi = [\,0.5 \,\cos \omega_e t \, \cdots \, \cos N\omega_e t$$
$$\sin \omega_e t \,\sin 2\omega_e t \, \cdots \, \sin N\omega_e t\,]^T \tag{7.26}$$

$$\psi_1 = [\,1 \,\cos \omega_e t \, \cdots \, \cos N\omega_e t$$
$$\sin \omega_e t \,\sin 2\omega_e t \, \cdots \, \sin N\omega_e t\,]^T \tag{7.27}$$

$$\omega_e = 2\pi f_s = \frac{2\pi}{T}. \tag{7.28}$$

The parameter ω_e is the fundamental angular frequency of the speed ripples and N can be chosen such that Fourier-series expansion covers the Nth-order harmonics of the fundamental frequency.

7.5 Implementation of Drive System

Figure 7.3 shows the overall speed-ripple minimization algorithm. During the transient state, the ILC is made inactive and i_{qs}^* is provided only by the PI speed controller output, i_{q0}^*. When steady state is reached, the ILC is applied and it provides the additional compensation term Δi_{qs}^* to i_{q0}^* so as to minimize the speed ripples. The performance of the drive system using the ILC algorithm is compared with the algorithm using only the PI controller. The sampling times for the controllers are: current controller 200 μs and the speed controller 800 μs. The gains of the PI speed controller are: $k_p = 0.035$, $k_i = 0.35$, while the gains of the PI current controllers are $k_p = 40$, $k_i = 600$. The learning gains are $\beta_l = 0.4$ and $\beta = 0.02$. The forgetting factor is $\alpha = 0.05$. Parameter N is chosen as 12, since only harmonics of order lower than or equal to 12 are considered in our study.

A photograph of a PMSM coupled with a DC generator used as a loading mechanism is shown in Fig. 7.4, whereas Fig. 7.5 shows the configuration of the experimental setup. Although there is a torque transducer coupled along the motor shaft, actually it is not used in our proposed algorithms, since we only need speed information instead of the torque signal. The motor parameters are listed in Table 7.1. The proposed control algorithm is realized in the DSP-based drive setup using the floating-point 60 MHz DSP TMS320C31. The proposed algorithm is implemented using a C-program. The currents of two phases are measured using Hall-effect devices and then converted into digital values using 16-bit A/D converters.

The performance evaluation of the proposed ILC control in suppressing torque ripples is presented in the next section.

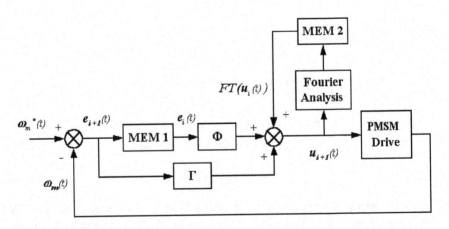

Fig. 7.2 Block diagram of the ILC algorithm implemented in the frequency domain

Speed Controller

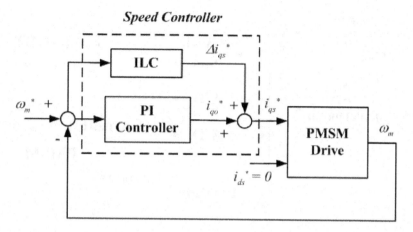

Fig. 7.3 Block diagram of the speed control loop used in the PMSM drive system

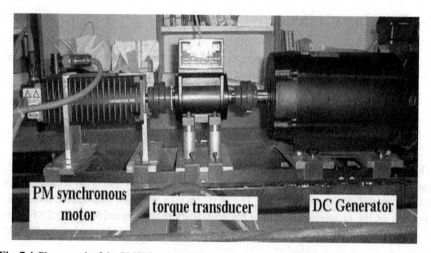

Fig. 7.4 Photograph of the PMSM and the DC generator used in experiments

Table 7.1 Motor Papameters

Rated power	1.64 kW
Rated speed	2000 rpm
Stator resistance	2.125 Ω
Stator inductance	11.6 mH
Magnet flux	0.387 Wb
Number of poles	6
Inertia	0.03 kgm^2

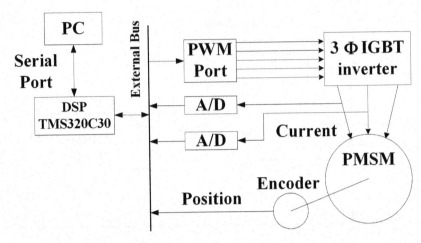

Fig. 7.5 Configuration of DSP-based experimental setup

7.6 Experimental Results and Discussions

To verify the effectiveness of the proposed ILC algorithms, experiments are carried out using the DSP-based PMSM drive system described in the previous section.

7.6.1 Experimental Results

The experiments were conducted under different operating conditions, with speeds ranging from 0.005 p.u. (10 rpm) to 0.05 p.u. (100 rpm) and load torques from 0.0 p.u. to 0.795 p.u. (6.20 N m). The performance criterion to evaluate the effectiveness of the proposed algorithm for speed-ripple minimization is the SRF. It is defined as the ratio of the peak-to-peak speed ripple to the rated speed of the PMSM

$$SRF = \frac{\omega_{pk-pk}}{\omega_{rated}} \times 100\%. \tag{7.29}$$

The SRF of the PMSM drive using the conventional PI speed controller alone is first determined. Subsequently, the proposed ILC controller that generates the compensation current is applied in parallel and the corresponding SRF is re-evaluated. Figures 7.6 to 7.11 each show the speed response in the time domain and in the corresponding frequency spectrum. For the purpose of clarity, we pick up only the 1st, 2,6th and 12th harmonics that are of interest and show them in the frequency spectrum. Figure 7.6 shows the speed waveform when the motor runs at 0.025 p.u. (50 rpm) under no external load torque and without ILC compensation. In this figure, a large speed oscillation can be observed with the corresponding SRF=0.65%. Figures 7.7 and 7.8 present the speed waveforms under the same working condition with ILC compensation algorithms implemented in the time domain and frequency domain, respectively. We can see that speed-ripple harmonics are reduced with the SRF=0.15% after applying the time-domain ILC algorithm. Further reduction is possible by using the frequency-domain learning method with the SRF reduced to 0.10%, comparing the 1st, 2nd and 6th harmonics in Fig. 7.7 with those in Fig. 7.8.

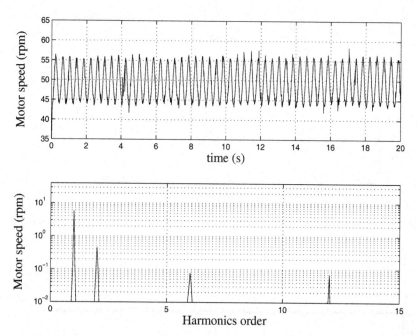

Fig. 7.6 The speed response without ILC compensation (ω_m = 50 rpm, T_l = 0.0 N m)

Figures 7.9 to 7.11 are arranged in the same sequence, and the operating conditions for these figures are $\omega_m = 0.025$ p.u. (50 rpm), $T_l = 0.795$ p.u. (6.20 N m). It can be seen that again both the proposed algorithms are able to compensate for the 1st and 2nd harmonic components to a greater extent. However, the reduction in the 6th and 12th components is not so considerable as the case without load. This is because when a motor operates under heavy loads, the DC generator is coupled and excited for the loading purpose, and the resultant torque pulsations induced from the load side that mainly consist of non-integer harmonics become greater. They are reflected back in speed ripples, which cannot be compensated by the ILC algorithms due to their non-integer multiples of the fundamental frequency nature.

The detailed SRFs under different working conditions are shown in Figs. 7.12 and 7.13. The former gives the SRFs when PMSM operates at different speeds: 0.005 p.u. (10 rpm), 0.025 p.u. (50 rpm) and 0.05 p.u. (100 rpm), under no external load without and with both ILC compensation algorithms. The latter shows the SRFs when PMSM operates at 0.025 p.u. (50 rpm) under different load torques: 0.0 p.u.,

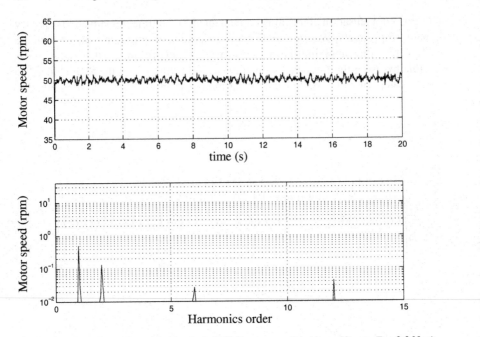

Fig. 7.7 The speed response with time-domain ILC compensation ($\omega_m = 50$ rpm, $T_l = 0.0$ N m)

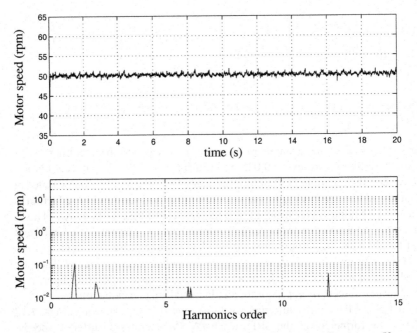

Fig. 7.8 The speed response with frequency-domain ILC compensation ($\omega_m = 50$ rpm, $T_l = 0.0$ N m)

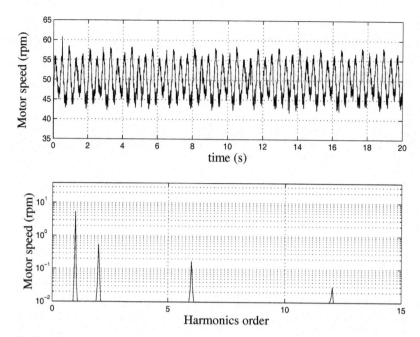

Fig. 7.9 The speed response without ILC compensation ($\omega_m = 50$ rpm, $T_l = 6.20$ N m)

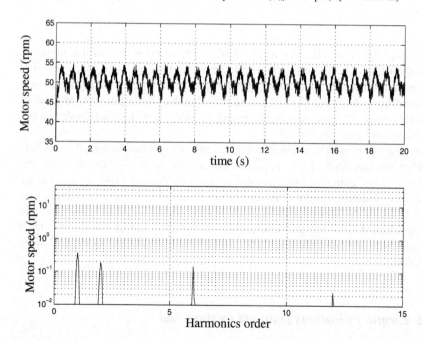

Fig. 7.10 The speed response with time-domain ILC compensation ($\omega_m = 50$ rpm, $T_l = 6.20$ N m)

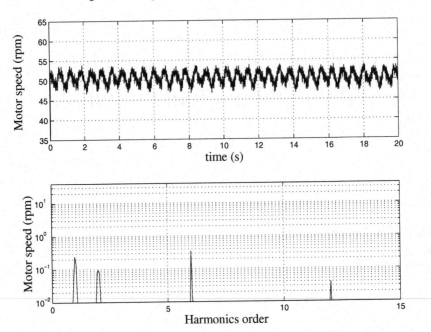

Fig. 7.11 The speed response with frequency-domain ILC compensation (ω_m = 50 rpm, T_l = 6.20 N m)

0.410 p.u. (3.2 Nm) and 0.795 p.u. (6.2 Nm), without and with both ILC compensation algorithms. According to the results, the effectiveness of the proposed ILC algorithms is verified in suppressing speed ripples under various steady-state operating conditions. As in Fig. 7.12, it can be seen that the SRF is relatively high when the motor runs at 50 rpm, since at this speed the peak-to-peak speed ripple reaches a maximum and so does SRF [25].

From the experimental results presented, apparently, it can be seen that the ILC algorithm implemented in the frequency domain by means of Fourier-series expansion is better than that implemented in the time domain. This is because the time-domain learning cannot eliminate errors totally due to the introduction of a forgetting factor [5]. However, there is still a disadvantage in the frequency-domain learning when it faces the problem caused by the existence of those non-integer harmonics induced from the load side, which will be explained in the following subsection.

7.6.2 Torque Pulsations Induced by the Load

It is necessary to distinguish those torque ripples caused by the loading mechanism from the ones induced by the PMSM, which are of our interest. For this purpose,

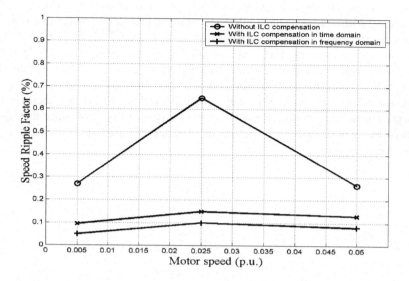

Fig. 7.12 SRFs of different speeds under no external load without and with ILC compensation

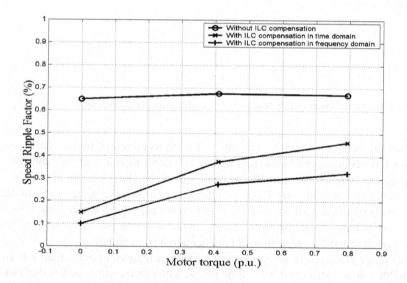

Fig. 7.13 SRFs of different load torques with speed at 50 rpm without and with ILC compensation

we decouple the PMSM from the DC machine, which is originally used as the load, excite the DC machine to work in the motoring mode at 50 rpm, and observe the torque signal by using a torque transducer. Figure 7.14 shows the frequency spectrum of the resultant torque pulsations induced by the DC machine alone. It can be seen that some non-integer harmonics are present in the torque waveforms, among which the 0.65th and 7.35th harmonics are dominant. In addition, energy distributed around the 7th and 8th harmonics are quite apparent. Figures 7.15 and 7.16 show the frequency spectrum of speed response without ILC compensation under a light load and a heavy load, respectively. In the former figure, we can observe that the integer multiple harmonics are the dominant components, while in the latter, non-integer harmonics, the 0.65th and 7.35th components become dominant. In fact, these components are the same dominant harmonics that appear in Fig. 7.14, since they are induced by the non-ideal load mechanism when the DC machine is excited for a heavy-load purpose.

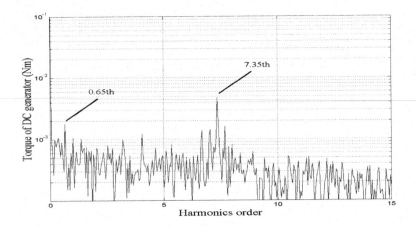

Fig. 7.14 Frequency spectrum of torque pulsations induced by the DC generator

It should be noted that the ILC control algorithms presented in this chapter are supposed to reduce speed harmonics, which are integer multiples of the fundamental frequency. This is because ILC is designed in the face of removing periodic disturbance in the input, and it has no effect on non-periodic disturbance. In our case, speed ripples caused by torque oscillations are periodic in nature, and most of these speed ripples share the same period. Thus, we define this time period as the basic period in the ILC algorithms implementation so as to remove most of these speed ripples. However, those non-integer harmonics induced from the load side are ripples that do not share that basic time period with others, which makes them impossible to eliminate by using ILC. To solve this problem, we can multiply the basic time period such that those non-integer harmonics become "integer" corresponding with the new period. However, this will prolong the converging time of the learning process, and such a method is non-effective in case of pseudo-harmonics, because it is impossible to find its time period.

This non-ideal loading mechanism, namely the appearance of non-integer torque harmonics further degrades the performance of the proposed controllers, particularly

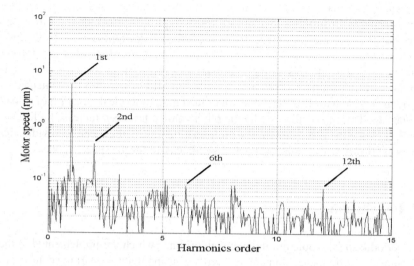

Fig. 7.15 Frequency spectrum of speed response without ILC compensation (ω_m = 50 rpm, T_l = 0.0 N m)

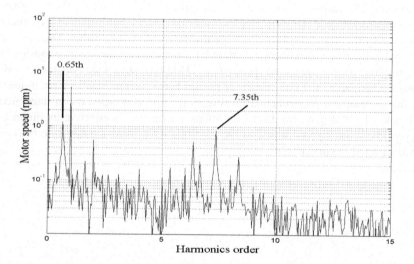

Fig. 7.16 Frequency spectrum of speed response without ILC compensation (ω_m = 50 rpm, T_l = 6.20 N m)

the frequency-domain learning method. Although a Fourier-based learning mechanism is capable of further reduction of speed ripples, it also induces a distorted approximation of these non-integer harmonics. Such distortion is caused by the mismatch between the time period of Fourier transformation and those of non-integer harmonics. As a result, this wrong information further induces the error accumulation during the learning process, which makes the ILC algorithm implemented in the frequency domain less desirable when facing the non-ideal situation of a load mechanism. This also explains why the 6th harmonic has been increased a little in Fig. 7.11, since the distortion of error information affects the local bandwidth and generates locally wrong control signals to compensate for the speed ripples.

7.7 Conclusion

Two periodic speed-ripple minimization algorithms, which are implemented in the ime domain and frequency domain, respectively, using iterative learning control are presented in this chapter. Learning control is intuitively an excellent selection for the speed-ripple minimization algorithm because of the periodic nature of torque and hence speed ripples. Moreover, the algorithm is simple to implement, can be added to any existing controller and does not require accurate knowledge of the motor parameters. The learning control algorithm implemented in the time domain guarantees minimization of the speed ripples to a certain extent, but further improvement is limited by the introduction of the forgetting factor in the algorithm. Therefore, a learning method in the frequency domain by means of Fourier-series expansion is designed to further suppress the speed ripples. However, such a method has a disadvantage when facing the non-integer or pseudo harmonics present in the non-ideal loading mechanism. Experimental investigations were conducted on an integrated DSP-based PMSM platform. Both algorithms are verified as effective in minimizing periodic speed ripples caused by the PMSM drive system and non-symmetrical construction of the motor.

Chapter 8
ILC for Electrical Drives: Application to a Switched Reluctance Motor

Abstract This chapter proposes the use of ILC in designing a torque controller for switched reluctance motors (SRM). The demanded motor torque is first distributed among the phases using a torque-sharing function. Following that, the phase torque references are converted to phase current references by a torque-to-current converter and the inner current control loop tracks the phase current references. SRM torque is a highly non-linear and coupled function of rotor position and phase current. Hence, the equivalent stator phase current for a given demanded torque can not be obtained analytically. The assumption of linear magnetization results in an invertible torque function. However, the equivalent current obtained using this torque function will lead to some error due to magnetic saturation. This error in the equivalent current will be periodic with rotor position for a given demanded torque. Hence, we propose to use *iterative learning* to learn this error to be added as a compensation for the equivalent current. Similarly, the current tracking for the non-linear and time-varying system is achieved by combining a simple P-type feedback controller with an ILC controller (CCL). The proposed scheme uses ILC to augment conventional techniques and hence, has better dynamic performance than a scheme using only ILC. Experimental results of the proposed scheme for an 8/6 pole, 1 hp SRM show very good average as well as instantaneous torque control.[1]

8.1 Introduction

A switched reluctance motor would have been the natural choice for variable-speed drive applications, had it not been for the large amount of torque ripples produced by it. With concentric windings on the stator and no windings or permanent magnet on its rotor, it is the simplest of all the types of electromagnetic motors. It is cheap to manufacture, robust and can operate under partial failure. Its power converter has no

---

[1] With IEEE permission to re-use "Indirect torque control of switched reluctance motors using iterative learning control," coauthored by Sahoo, S.K., Panda, S.K., Xu, J.CX., IEEE Transactions on Power Electronics, Vol. 20, no.1, pp.200–208, 2005

chance for shoot-through faults. With the seminal paper [69], there has been a surge in research activities to avail the numerous advantages of SRM in a variable-speed drive application.

SRM torque can be expressed as the rate of change of co-energy (W_c) with rotor position:

$$T(i,\theta) = \frac{\partial W_c}{\partial \theta} = \frac{\partial}{\partial \theta} \int_0^i \psi \, di. \tag{8.1}$$

SRM has to operate in deep magnetic saturation to achieve a torque density comparable with other types of motors. The double saliency of the motor gives rise to flux fringing when the stator and rotor poles are close to each other. The combined effect of flux fringing and saturation makes both flux linkage and torque a highly non-linear and coupled functions of stator current and rotor position. The torque function is not invertible and therefore it is not possible to find the equivalent stator phase currents for a given desired torque. When saturation and flux fringing are ignored, the phase inductance has a trapezoidal profile. Hence, the phase torque expression in the increasing inductance region would be:

$$T(i,\theta) = \frac{1}{2} i^2 \frac{dL}{d\theta} \tag{8.2}$$

and the equivalent reference phase current for a given phase torque reference would be a function of θ:

$$i_{ref}(\theta) = \sqrt{\frac{2T_{ref}(\theta)}{\frac{dL}{d\theta}}}. \tag{8.3}$$

However, at higher torque levels when the motor enters into magnetic saturation, the above approach leads to an error in average output torque value and results in a large amount of ripples.

The other source of error in torque control is inaccurate tracking of the phase current references by the inner-loop current controllers. The discrete nature of phase current references in addition to varying phase inductance with both rotor position and phase current results in a non-linear time-varying tracking control problem. Popular non-linear controllers like variable structure with SMC or feedback linearization requires either the knowledge of structured uncertainties, or an accurate model of the plant. As the accurate modelling of SR motors is extremely difficult, robustness of the controller to modelling error is highly desirable.

Iterative learning control was used for robots [4], where the operation was periodic and the system states matched the desired states at the start of each period. ILC learns the desired control input by iteratively updating the input so as to reduce the tracking error, and does not require a detailed plant model. For SRM, the phases are excited alternately. At any time the demanded motor torque is distributed between two phases, while the other phases are switched off. For a given demanded torque and constant speed, the torque reference of each phase is periodic along the rotor position axis. Also, the desired phase torque as well as the actual phase torque are zero at the start of each period. Hence, we propose to use ILC to design a torque controller for SRM. For actual motor operation, neither torque nor speed remains constant. Keeping this in mind, we propose to use ILC as a plug-in controller to an existing conventional controller. The conventional controllers contribute most of the total control effort during transient periods. Hence, torque control during a transient period is not quite accurate. After the motor reaches constant speed and constant torque operation, ILC converges and provides accurate torque control. Accurate and ripple-free torque production is more critical for constant-speed and torque operations. Thus, the proposed scheme is useful for practical operation.

We use iterative learning in both the torque to current conversion and current controller design. The equivalent phase current reference for a given phase torque reference is thought of as having two parts. We make use of the simple invertible relationship between current and torque based on an idealized trapezoidal inductance profile to determine the major part of the current reference. An iteratively learned compensation current is then added to the first part to obtain the overall current reference. The compensation current takes care of all types of non-linearities in the SRM. Similarly, an iterative-learning-based current controller is added to a simple P-type feedback current controller to ensure accurate current tracking. Instead of using a look-up table for torque estimation, an analytical torque model is derived from the flux-linkage model. The model parameters are determined from least squares error fitting of static torque measurements.

The chapter is structured as follows. Section 8.2 gives a brief review of earlier studies. Section 8.3 discusses the cascaded torque control structure, with details on the three subunits, described in Sects. 8.3.1, 8.3.2 and 8.3.3. The analytical torque estimator is described in Sect. 8.3.4. Section 8.4 contains the experimental figures and discussions. Section 8.5 concludes the chapter.

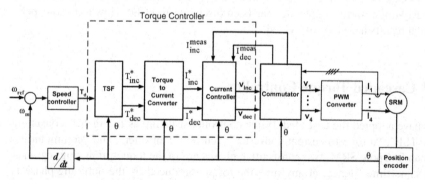

Fig. 8.1 Closed-loop speed control of SRM with an indirect torque controller

8.2 Review of Earlier Studies

Husain [50] has done an extensive survey on various torque-ripple minimization techniques proposed in the last decade. Major contributions in this line are reported in [11, 36, 65, 92, 104, 106, 120, 132, 141, 142]. With the aim of maximizing efficiency and the speed range, Husain proposed a hybrid controller based on the strengths of some of the earlier works. The phase torque reference is converted to a current reference by using a look-up table. Hysteresis controller implemented on an analog circuit is used during the period of active current control.

Another method for mapping of the non-linear relationship is the use of artificial neural networks [92, 104]. The demanded torque and rotor position are taken as input to obtain the equivalent stator phase current as the output of the network. Neural networks require much less storage space as only the synaptic weights and biases have to be stored. However, they involve a large amount of on-line computation as compared to the look-up tables.

A hysteresis controller could ensure good tracking for a non-linear system, if implemented on an analog system. On digital systems with finite sampling frequency, such a large gain controller will lead to a large amount of current ripple. Therefore, hysteresis current control is not appropriate for SRM, which requires a digital controller. The feedback linearization scheme proposed in [11] used a trapezoidal phase inductance profile that ignores magnetic saturation and hence is not accurate. A nonlinear IMC controller for both torque and current control of SRM was proposed in [36], which uses analytical flux-linkage and torque models, and plant-model error feedback through a filter to compensate for modelling inaccuracies. This scheme involves excessive on-line computation for state transformations. Application of ILC in designing a torque controller has been discussed in [109]. However, no experimental results have been provided.

8.3 Cascaded Torque Controller

We have adopted the cascaded torque controller structure as in Fig. 8.1, consisting of: "TSF", "torque to current converter", "current controller", and "commutator". For an 8/6 pole SRM, two consecutive phases are in the toque-producing region at the same time. Hence, at any time the torque produced by the outgoing phase is slowly reduced and that of the incoming phase slowly increased. Based on the rotor position, a pre-determined TSF, as shown in Fig. 8.2, distributes the demanded torque into the increasing component (T_{inc}) and a decreasing component (T_{dec}). The torque to current conversion block then converts them to I_{inc} and I_{dec}, respectively. The closed-loop current controller subsequently determines the voltages V_{inc} and V_{dec} to be applied to the increasing and decreasing phases, respectively. The commutator block assigns the role of increasing and decreasing phase voltages to all the four phases in sequence.

8.3.1 The TSF

As instantaneous commutation of phase currents is not possible, the TSF results in smoothly varying current references. As many TSFs can achieve this; additionally, the motor efficiency and operating speed range should be optimized. In this chapter, a TSF containing a constant and a cubic segment as shown in Fig. 8.2, is used. For an 8/6 pole SRM, each phase conducts for 30°(mechanical). Hence, we consider the phase current to be increasing for $0° - 15°$ and decreasing for $15° - 30°$. The T_{inc} has a *zero-torque*, a *rising-torque*, and a *full-torque* segment. The cubic segment contains one phase current rising and the other phase falling, and is referred to as the phase *overlap period*. The cubic polynomial $f(\theta)$ is:

$$f(\theta) = A + B(\theta - \theta_{on}) + C(\theta - \theta_{on})^2 + D(\theta - \theta_{on})^3,$$

where A, B, C, D are constants, chosen so as to satisfy constraints as will be stated in (8.7); θ is the rotor position, θ_{on} is the on-angle and θ_v is the "overlap period". The T_{inc} and T_{dec} are defined as

$$T_{inc} = \begin{cases} 0 & \text{for } 0 \leq \theta < \theta_{on} \\ f(\theta) & \text{for } \theta_{on} \leq \theta < \theta_{on} + \theta_v \\ T_{ref} & \text{for } \theta_{on} + \theta_v \leq \theta < 15° \end{cases} \tag{8.4}$$

and

$$T_{dec} = T_{ref} - T_{inc},$$

respectively. The constraints for the rising segment are defined by:

$$f(\theta) = \begin{cases} 0 & \text{at } \theta = \theta_{on} \\ T_{ref} & \text{at } \theta = \theta_{on} + \theta_v \end{cases} \tag{8.5}$$

and

$$\frac{df(\theta)}{d\theta} = \begin{cases} 0 & \text{at } \theta = \theta_{on} \\ 0 & \text{at } \theta = \theta_{on} + \theta_v. \end{cases} \tag{8.6}$$

With these constraints (8.5) and (8.6), the various constants of the polynomial $f(\theta)$ can be derived as,

$$A = 0; B = 0; C = \frac{3T_{ref}}{\theta_v^2}; D = -\frac{2T_{ref}}{\theta_v^3}. \tag{8.7}$$

The overlap period θ_v should be small to ensure that the torque-producing responsibility is transferred as fast as possible leading to better efficiency. However, a small θ_v would reduce the operating speed range. With this in mind, θ_v can be determined for a given system. The on-angle θ_{on} is chosen so that commutation occurs around the critical angle θ_c, as defined in [36]. This commutation method produces active current control throughout the conduction period.

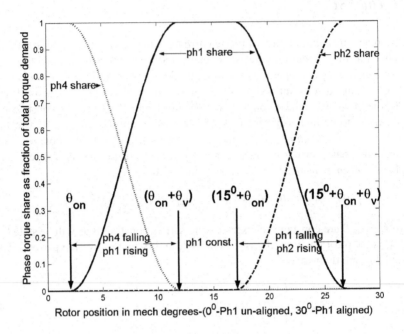

Fig. 8.2 Cubic torque-sharing function

8.3.2 Proposed Torque to Current Conversion Scheme

When the motor operates in the linear magnetic region, the phase inductance can be simplified as a trapezoidal function of rotor position only. At higher torque levels, the motor enters into magnetic saturation, and the phase inductance will be a function of both i and θ. With this understanding, we find the current reference for a given torque level as a sum of two parts, as shown in Fig. 8.3. Part 1 is obtained by assuming the SRM inductance profile has an ideal trapezoidal shape. This assumption leads to a non-coupled torque relationship that can be inverted as in (8.3).

Part 2 is an iteratively learned compensation current at each rotor position, which is dependent on the degree of saturation. The correctness of the current reference is checked by converting it to torque through the torque estimator and comparing it with the torque reference. Any difference between this estimated torque and the demanded torque is stored in memory against rotor positions. Based on this error, the compensation current is updated from iteration to iteration according to the learning law given in (8.8), until the error is within a pre-specified limit. The amount of compensation current is dependent on the magnitude of the motor torque.

$$\triangle I_{ilc}(z,k) = \triangle I_{ilc}(z-1,k) + \beta_1 \times T_{err}(z-1,k) \tag{8.8}$$

$$T_{err}(z-1,k) = T^* - T_{est}(z-1,k), \tag{8.9}$$

where $\triangle I_{IL}(z,k)$ and $\triangle I_{IL}(z-1,k)$ are the iteratively learned compensation currents at the kth position interval for zth and $(z-1)$th iterations, respectively; T^* is either the T_{inc} or T_{dec} and $T_{est}(z-1,k)$ is the corresponding estimated torque for the current reference at kth position interval for $(z-1)$th interval respectively. The learning gain β_1 is chosen from the following inequality that ensures ILC convergence,

$$|1 - \beta_1 G_1| < 1 \Rightarrow 0 < \beta_1 < \frac{2}{G_1},$$

where $G_1 = \frac{\partial T}{\partial i}$. As the value of G_1 is used only for determining the maximum

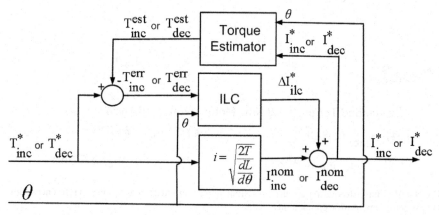

Fig. 8.3 Proposed torque to current conversion with ILC compensation

allowed learning gain, it can be obtained from (8.2) as $i\frac{dL}{d\theta}$. For the SRM used in our experiment, the highest value of $\frac{dL}{d\theta}$ is 0.14 in the unsaturated case. Hence, with a maximum current value of $12A$, range of β_1 for ILC convergence,

$$0 < \beta_1 < 1.19. \tag{8.10}$$

The learning gain range obtained above is conservative as the actual G_1 will be less when motor goes into saturation. We have used a value of 1.0 in our experiment. Figure 8.4 shows the increasing phase current reference (I_{inc}) and decreasing phase current reference (I_{dec}), along with the corresponding ILC-based compensation currents $\triangle I_{ilc,inc}$ and $\triangle I_{ilc,dec}$; for a demanded torque of 1.2 N m. As can be seen, the compensation current for the increasing phase is small in magnitude. This is because of the fact that the incoming phase is near the unaligned position and has not entered much into saturation. As the rotor pole approaches the aligned position, the pole tips enter into deep saturation and hence more compensation current is required. This is reflected in the compensation current for the decreasing phase.

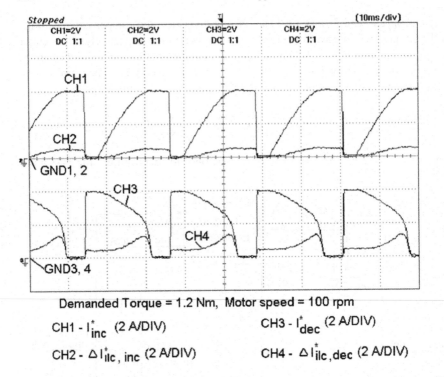

Demanded Torque = 1.2 Nm, Motor speed = 100 rpm

CH1 - I_{inc}^* (2 A/DIV)   CH3 - I_{dec}^* (2 A/DIV)

CH2 - $\Delta I_{ilc, inc}^*$ (2 A/DIV)   CH4 - $\Delta I_{ilc, dec}^*$ (2 A/DIV)

Fig. 8.4 ILC-based torque to current conversion: increasing and decreasing current references with ILC-based compensation component

8.3.3 ILC-based Current Controller

It is common knowledge that a feedforward control scheme improves the trajectory-tracking performance. With exact knowledge of the plant model, the inverse model, if possible, can be used as a feedforward controller to achieve perfect tracking. As mentioned earlier, it is difficult to obtain an accurate model for SRM. Hence, a model-dependent feedforward controller would make SRM less attractive. To solve this problem, we have proposed a novel *iterative-learning*-based current controller, shown in Fig. 8.5. The proposed current controller, consisting of a simple P-type feedback controller and an ILC block as the feedforward controller, as described below.

$$u = u_{fb} + u_{ilc}. \tag{8.11}$$

For constant speed and constant torque, the phase current reference can be tracked by applying a position-dependent voltage, which is repetitive in nature. Thus, the ILC has to learn this required voltage as a function of rotor position. We divide one learning period into smaller intervals of $0.1°$ each, such that required voltage profile can be approximated to be of constant value over each interval. During each sampling period, the error in current is calculated and stored in memory against the

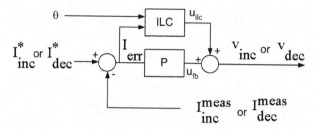

Fig. 8.5 Block diagram for the proposed current controller: P-type feedback control with ILC compensation

corresponding position interval, along with the voltage applied to the phase winding. One period is equivalent to one learning iteration. During the next period, the new voltage for each position interval is calculated as per the learning law,

$$u_{ilc}(z, k) = u(z - 1, k) + \beta_2 \times i_{err}(z - 1, k + 1), \quad (8.12)$$

where $u_{ilc}(z, k)$ is the ILC output during the zth learning iteration and at kth position interval, $u(z - 1, k)$ is the applied voltage during the $(z - 1)$th learning iteration and at the kth position interval, β_2 is the learning gain, $i_{err}(z - 1, k + 1)$ is the filtered current error during the $(z - 1)$th learning iteration and at $(k + 1)$th position interval. To account for the calculation delay in a digital system, the error at one position interval ahead is used in updating the voltage at the present position interval.

For our proposed learning law, the range for β_2 for ILC convergence is determined from the following inequality,

$$|1 - \beta_2 G_2| < 1 \Rightarrow 0 < \beta_2 < \frac{2}{G_2}, \quad (8.13)$$

where G_2 is defined in the sampled-data state-space equation for the system

$$i(k + 1) = F_2(i(k), \theta(k))i(k) + G_2(i(k), \theta(k))u(k), \quad (8.14)$$

where $i(k)$, $u(k)$ are phase current and phase voltage, respectively, at sampling instant k. The value of G_2 can be obtained from their corresponding continuous time-domain values, which depends on the stator phase resistance and incremental inductance. As we need to find the maximum value of G_2, we can use the minimum phase inductance at the unaligned position.

The square of the current error over each iteration is summed and compared with an acceptable error criteria. Based on this comparison, it is decided to continue or stop learning during the next iteration. Thus, the ILC is automatically started and stopped in the actual system. Figure 8.6, shows the performance of only a P-type current controller, for a demanded torque of 1.2 N m and motor speed of 100 rpm. As can be seen, there is a substantial current tracking error. For a constant proportional gain, this error will be different for other torque levels as well as motor

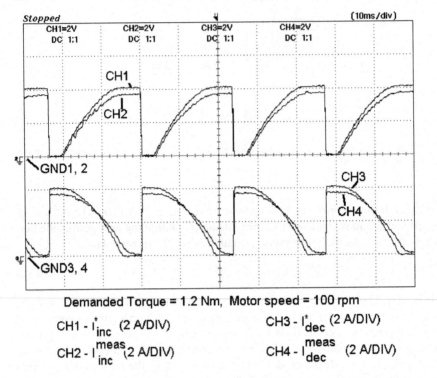

Fig. 8.6 Current tracking with P-type current controller : Increasing and decreasing current references with corresponding current feedbacks; before ILC is activated

speeds. In Fig. 8.7, with the ILC compensation voltage added to the P-type current controller output, the reference current and measured current are indistinguishable. Thus, perfect current tacking is possible in the proposed current controller. The current reference has changed as compared to that in Fig.8.6 because of ILC-based compensation in torque to current conversion.

8.3.4 Analytical Torque Estimator

For torque estimation, we use an analytical expression based on flux-linkage modelling. Though there are numerous proposed flux-linkage models, we find the model reported in [136] to be intuitive as well as compact for flux-linkage modelling. This model captures the flux linkage quite accurately and is based on the physical properties of the SRM.

$$\psi(i, \theta) = a_1(\theta)(1 - e^{(a_2(\theta)i)}) + a_3(\theta)i, \qquad (8.15)$$

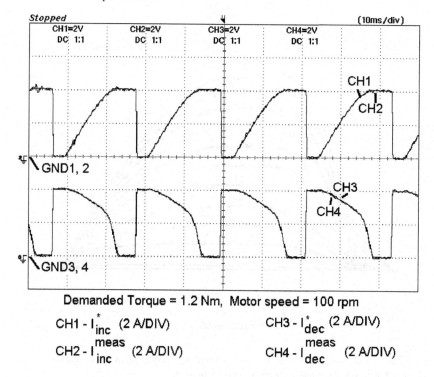

Demanded Torque = 1.2 Nm, Motor speed = 100 rpm

CH1 - I^*_{inc} (2 A/DIV)   CH3 - I^*_{dec} (2 A/DIV)

CH2 - I^{meas}_{inc} (2 A/DIV)   CH4 - I^{meas}_{dec} (2 A/DIV)

Fig. 8.7 Current tacking with ILC compensation to P-type current controller : Increasing and decreasing current references with corresponding current feedbacks; after ILC is activated

where $a_1(\theta), a_2(\theta)$ and $a_3(\theta)$ have the following physical meanings: $a_1(\theta)$ is the saturated flux linkage, $a_2(\theta)$ captures the severity of saturation, and $a_3(\theta)$ is the incremental inductance at high currents. We have used three fifth-order polynomials in θ to approximate these coefficient functions. The expression for torque in terms of phase current and rotor position is then derived from the above flux-linkage model by expanding (8.1).

$$T = \{i + \frac{1}{a_2}(1 - e^{a_2 i})\}\frac{da_1}{d\theta}$$
$$-\{\frac{a_1}{a_2^2}(1 - e^{a_2 i}) + \frac{a_1 i}{a_2}e^{a_2 i}\}\frac{da_2}{d\theta} + \frac{1}{2}i^2\frac{da_3}{d\theta}. \qquad (8.16)$$

However, unlike [136] where the model parameters were obtained from curve fitting with the flux-linkage data, we have used the static torque data for SRM. The resultant torque matching is excellent, as shown in Fig. 8.8. The model-based estimator obtained through least squares is less prone to measurement noise than the look-up-table-based method. Also, the computation involved in such a model-based estimator is less compared to an artificial neural network for the same purpose.

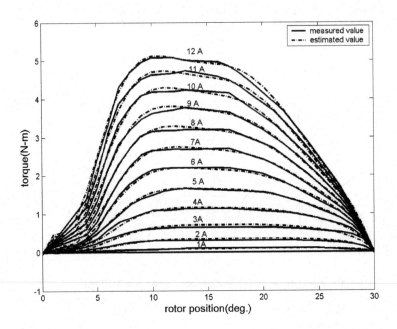

Fig. 8.8 Matching of measured static torque with the analytical model predicted torque for the 8/6 pole SRM, for different stator phase currents and rotor positions

8.4 Experimental Validation of the Proposed Torque Controller

We have implemented the proposed torque controller on our experimental platform, consisting of an 8/6 pole, 4-ph, 1-hp, 4000 rpm SRM and a dSPACE-DS1104 DSP controller. The sampling time is $150 \mu s$. A four-phase asymmetric bridge converter is used for supplying the SRM. A DC machine is used for loading the SRM.

The proposed indirect torque controller augments the conventional schemes for torque to current conversion and current controller; with ILC as an add-on controller. To show the contribution of ILC towards the performance enhancement of the proposed torque controller, we have shown the toque control performance both before and after ILC compensation. The torque error in the conventional scheme is due to the error in both torque to current conversion and current tracking. The error in the torque to current conversion is more at higher levels of demanded torque when the motor goes deep into saturation. This results in a larger average torque error. A P-type feedback current controller alone leads to more current tracking error at higher motor speeds. However, accurate torque to current conversion and current tracking by the inner-loop current controller should ensure accurate tracking of increasing and decreasing torque references.

Figure 8.9 shows the increasing (T_{inc}) and decreasing (T_{dec}) phase torque references, and corresponding torque estimations ($T_{inc,est}$) and ($T_{dec,est}$) for a demanded

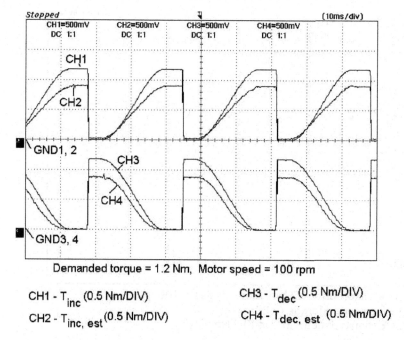

Demanded torque = 1.2 Nm, Motor speed = 100 rpm

CH1 - T_{inc} (0.5 Nm/DIV)   CH3 - T_{dec} (0.5 Nm/DIV)

CH2 - $T_{inc, est}$ (0.5 Nm/DIV)   CH4 - $T_{dec, est}$ (0.5 Nm/DIV)

Fig. 8.9 Experimental result: conventional torque controller performance without ILC compensation; increasing and decreasing torque references with corresponding estimated torque values

torque of 1.2 N m and motor speed of 100 rpm. Only the conventional schemes for torque to current conversion and current controller were used. There is substantial phase torque tracking error that results in both average torque error and a large number of torque ripples, as can be seen in Fig. 8.10. Figure 8.11 shows that with ILC compensation added, there is very good tracking for both the increasing and decreasing torque references, for the same demand torque and motor speed. This reduces the average torque error as well as the torque ripples, as can be seen in Fig. 8.12. The convergence of the ILC compensation scheme can be seen from Figs. 8.13 and 8.14, which show the estimated torque before and after the ILC compensation in one figure. As can be seen, the average torque error before ILC compensation is about 20% and 25% for motor speeds of 150 rpm and 250 rpm, respectively, whereas, the average torque errors are eliminated with activation of ILC at the same time, torque ripples are reduced from about 15% to about 5%. For verification of the proposed scheme over the complete torque range, we have carried out the experiment for 0.6 N m and 2.0 N m to cover the complete torque range. They are repeated for two different speeds in the low-speed range. Figure 8.15 shows the demanded torque and estimated torque, with demanded torque being 0.6 N m, and motor running at 75 rpm. Before the ILC compensation is added, there was an average torque error of about 10%, and torque ripple of 20%. Once ILC has converged after about 5 cycles, the average torque error is almost eliminated and torque ripples are reduced to less than 10%. Figure 8.16 shows the torque control performance when speed

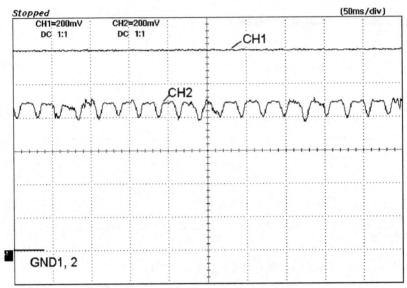

Demanded torque = 1.2 Nm, Motor speed = 100 rpm
CH1 - Demanded torque (0.2 Nm/DIV)
CH2 - Estimated motor torque (0.2 Nm/DIV)

Fig. 8.10 Experimental result: conventional torque controller performance without ILC compensation; demanded torque (CH1) and estimated motor torque (CH2)

is increased from 75 rpm to 150 rpm. The average torque error before ILC is activated is about 16% and is almost eliminated after ILC has converged. The torque ripples have been reduced from 20% to less than 10%. Figure 8.17 and 8.18 show the results for a demanded torque of 2.0 N m which is slightly above the rated torque of 1.78 N m, for motor speeds of 150 rpm and 300 rpm, respectively. The conventional schemes have an average torque error of 25% and 35%, respectively, which are almost eliminated after ILC compensation is added. The torque ripples are reduced from about 15% with the conventional scheme to less than 10% with the ILC compensation. These results have demonstrated the effectiveness of ILC in achieving better average torque control as well as reduced torque ripples, over the whole range of motor torque for low speeds of operation. It can also be noted that ILC takes about 5 cycles to converge.

Next, we would like to test the complete torque controller under normal operation, *e.g.* the motor is started from rest and accelerated to a demanded speed, or the motor load is changed in steps. Figure 8.19 shows the case when full motor torque is applied to accelerate the motor from rest. The motor speed increases to about 300 rpm. As can be seen, the average torque error reduces at the same rate as the ILC convergence. This shows the dynamic performance of the proposed torque control scheme. In Fig. 8.20, the demanded torque is reduced from 2 N m to 1 N m with the motor running at 300 rpm. As can be seen in the figure, the torque controller

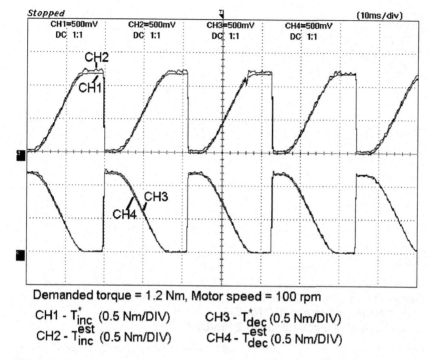

Demanded torque = 1.2 Nm, Motor speed = 100 rpm

CH1 - T^*_{inc} (0.5 Nm/DIV)   CH3 - T^*_{dec} (0.5 Nm/DIV)

CH2 - T^{est}_{inc} (0.5 Nm/DIV)   CH4 - T^{est}_{dec} (0.5 Nm/DIV)

Fig. 8.11 Experimental result: proposed ILC-based torque controller: increasing and decreasing torque references with corresponding estimated torque values

achieves the new demanded torque level accurately and quickly. The torque ripples are within 10%, except for the short transient period.

The experiment has been limited to within 10% of the rated speed of the motor. In the ILC design, the rotor position corresponding to one iteration is divided into smaller intervals that store the error and compensation values. During each iteration, all these data should be updated. At some higher speeds, when the angular displacement for one program execution period is more than the width of one ILC position interval, it may lead to performance degradation as data for all positions will not be updated in each learning cycle. Hence, as understood, high sampling frequency is desirable for high torque control performance over a large speed range. However, as torque ripples are critical at low speeds for many low speed applications; the proposed scheme is quite attractive for low speed, direct-drive applications.

8.5 Conclusion

In this chapter, we have proposed to use *iterative learning* in developing a torque controller for SRM that does not require detailed SRM modelling. The torque to

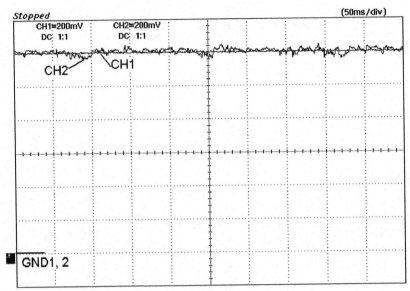

Demanded torque = 1.2 Nm, Motor speed = 100 rpm
CH1 - Demanded torque (0.2 Nm/DIV)
CH2 - Estimated motor torque (0.2 Nm/DIV)

Fig. 8.12 Experimental result: proposed ILC based torque controller; demanded torque (CH1) and estimated motor torque (CH2)

current conversion method adds a compensation current to a conventional method for magnetic saturation. It does not require as much on-line memory as a look-up table, or as much computation as an artificial neural-network-based method. Hence, it is easy for real-time implementation. A flux model based analytical torque estimator was used for estimating the torque from the phase currents and rotor position. An ILC based compensation voltage is added to a P-type feedback current controller for accurate current tracking. The combined cascaded torque controller has been experimentally validated. It can achieve accurate average torque control while minimizing the torque ripples. The validity of the proposed scheme at higher speeds will be the subject of future investigations.

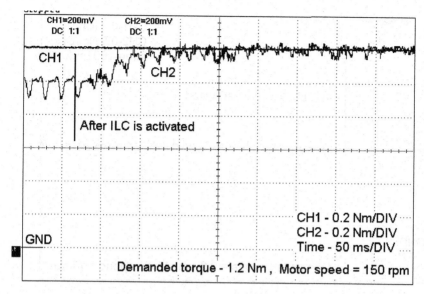

Fig. 8.13 Experimental result: proposed ILC based torque controller; demanded torque (CH1)= 1.2 N m, estimated motor torque (CH2), motor speed = 150 rpm

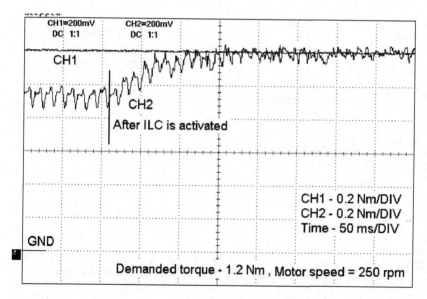

Fig. 8.14 Experimental result: proposed ILC-based torque controller; demanded torque (CH1) = 1.2 N m, estimated motor torque (CH2), motor speed = 250 rpm

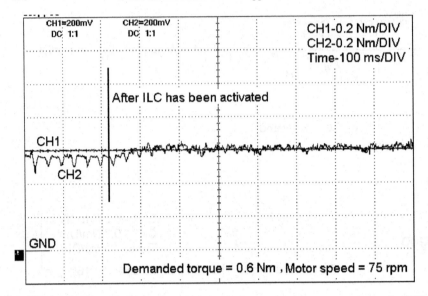

Fig. 8.15 Experimental result: proposed ILC-based torque controller; demanded torque (CH1) = 0.6 N m, estimated motor torque (CH2), motor speed = 75 rpm

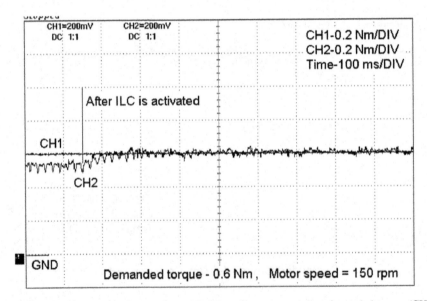

Fig. 8.16 Experimental result: proposed ILC-based torque controller; demanded torque (CH1)= 0.6 N m, estimated motor torque (CH2), motor speed = 150 rpm

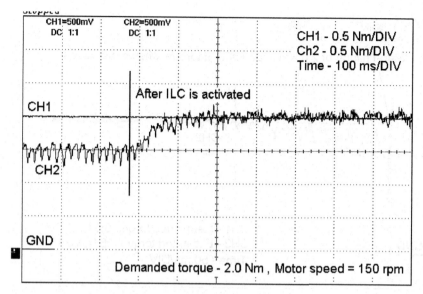

Fig. 8.17 Experimental result: proposed ILC-based torque controller; demanded torque (CH1) = 2.0 N m, estimated motor torque (CH2), motor speed = 150 rpm

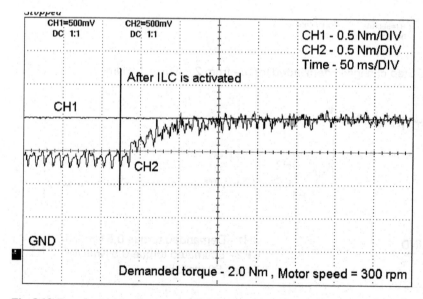

Fig. 8.18 Experimental result: proposed ILC-based torque controller; demanded torque (CH1) = 2.0 N m, estimated motor torque (CH2), motor speed = 300 rpm

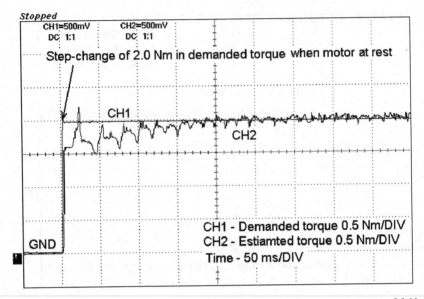

Fig. 8.19 Experimental result: proposed ILC-based torque controller; step change of 2 N m in demanded torque (CH1) when motor was at rest, estimated motor torque (CH2)

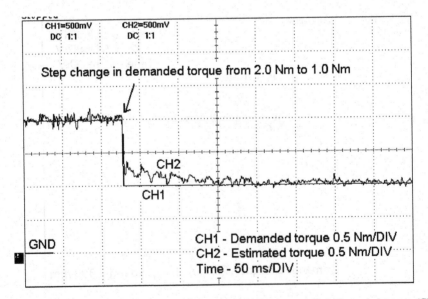

Fig. 8.20 Experimental result: proposed ILC-based torque controller; demanded torque (CH1) reduced from 2 N m to 1 N m when motor was moving at 150 rpm, estimated motor torque (CH2)

Chapter 9
Optimal Tuning of PID Controllers Using Iterative Learning Approach

Abstract The proportional-integral-derivative (PID) controller is the predominant industrial controller that constitutes more than 90% of feedback loops. Time-domain performance of PID, including peak overshoot, settling time and rise time, is directly dependent on the PID parameters. In this work we propose an iterative learning tuning method (ILT) – an optimal tuning method for PID parameters by means of iterative learning. PID parameters are updated whenever the same control task is repeated. The first novel property of the new tuning method is that the time-domain performance or requirements can be incorporated directly into the objective function to be minimized. The second novel property is that the optimal tuning does not require as much plant-model knowledge as other PID tuning methods. The new tuning method is essentially applicable to any plants that are stabilizable by PID controllers. The third novel property is that any existing PID auto-tuning methods can be used to provide the initial setting of PID parameters, and the iterative learning process can achieve a better PID controller. The fourth novel property is that the iterative learning of PID parameters can be applied straightforwardly to discrete-time or sampled-data systems, in contrast to existing PID auto-tuning methods that are dedicated to continuous-time plants. In this chapter, we further exploit efficient search methods for the optimal tuning of PID parameters. Through theoretical analysis, comprehensive investigations on benchmarking examples, and real-time experiments on the level control of a coupled-tank system, the effectiveness of the proposed method is validated.[1]

9.1 Introduction

Among all the known controllers, the proportional-integral-derivative (PID) controllers are always the first choice for industrial control processes owing to the

---

[1] With SICE permission to re-use the paper "Optimal tuning of PID parameters using iterative learning approach," coauthored by Xu, J.X., Huang, D.Q., Srinivasan, P., SICE Journal of Control, Measurement, and System Integration, Vol. 1, no.2, pp.143–154, 2008.

simple structure, robust performance, and balanced control functionality under a wide range of operating conditions. However, the exact workings and mathematics behind PID methods vary with different industrial users. Tuning PID parameters (gains) remains a challenging issue and directly determines the effectiveness of PID controllers [56, 93, 8].

To address the PID design issue, much effort has been invested in developing systematic auto-tuning methods. These methods can be divided into three categories, where the classification is based on the availability of a plant model and model type, (i) non-parametric model methods; (ii) parametric model methods; (iii) model-free methods.

The non-parametric model methods use partial modelling information, usually including the steady-state model and critical-frequency points. These methods are more suitable for closed-loop tuning and are applied without the need for extensive prior plant information [56]. The relay feedback tuning method [6, 140, 74, 79, 125] is a representative method of the first category.

The parametric model methods require a linear model of the plant – either a transfer function matrix or state-space model. To obtain such a model, standard off-line or on-line identification methods are often employed to acquire the model data. Thus, parametric model methods are more suitable for off-line PID tuning [9]. When the plant model is known with a parametric structure, optimal design methods can be applied [52, 68]. As for PID parameter tuning, it can be formulated as the minimization of an objective function with possible design specifications, such as the nominal performance, minimum input energy, robust stability, operational constraints, *etc*.

In model-free methods, no model or any particular points of the plant are identified. Three typical tuning methods are unfalsified control [108], iterative feedback tuning [73] and extremum seeking [61]. In [108], input–output data is used to determine whether a set of PID parameters meets performance specifications and these PID parameters are updated by an adaptive law based on whether or not the controller falsifies a given criterion. In [73], the PID controller is updated through minimizing an objective function that evaluates the closed-loop performance and estimates the system gradient (iterative feedback tuning or IFT method). In [61], adaptive updating is conducted to tune PID parameters such that the output of the cost function reaches a local minimum or local maximum (extremum seeking or ES method).

In practice, when the plant model is partially unknown, it would be difficult to compute PID parameters even if the relationship between transient specifications and PID parameters can be derived. In many existing PID tuning methods, whether model-free or model-based, test signals will have to be injected into the plant in order to find certain relevant information for setting controller parameters. This testing process may, however, be unacceptable in many real-time control tasks. On the other hand, many control tasks are carried out repeatedly, such as in batch processes. The first objective of this work is to explore the possibility of fully utilizing the task repetitiveness property, consequently providing a learning approach to improve PID controllers through iteratively tuning parameters when the transient behavior is the main concern.

In most learning methods, including neural learning and iterative learning [14], the process Jacobian or gradient plays the key role by providing the greatest descending direction for the learning mechanism to update inputs. The convergence property of these learning methods is solely dependent on the availability of the current information on the gradient. The gradient between the transient control specifications and PID parameters, however, may not be available if the plant model is unknown or partially unknown. Further, the gradient is a function of the PID parameters, thus the magnitude and even the sign may vary. The most difficult scenario is when we do not know the sign changes *a priori*. In such circumstances, traditional learning methods cannot achieve learning convergence. The second objective of this work is to extend the iterative learning approach to deal with the unknown gradient problem for PID parameter tuning.

It should be noted that in many industrial control problems, such as in the process industry, the plant is stable in a wide operation range under closed-loop PID control, and the major concern for a PID tuning is the transient behaviors either in the time domain, such as peak overshoot, rise time, settling time, or in the frequency domain, such as bandwidth, damping ratio and undamped natural frequency. From the control-engineering point of view, it is a very challenging task to directly address the transient performance, in comparison with the stability issues, by means of tuning control parameters. Even for a lower-order linear time-invariant plant under PID, the transient performance indices such as overshoot could be highly non-linear in PID parameters and an analytical inverse mapping from overshoot to the PID parameters may not exist. In other words, from the control specification on overshoot we are unable to decide the PID parameters analytically. The third objective of this work is to link these transient specifications with PID parameters and give a systematic tuning method.

Another issue is concerned with the redundancy in PID parameter tuning when only one or two transient specifications are required, for instance when only overshoot is specified, or only the integrated absolute error is to be minimized. In order to fully utilize the extra degrees of freedom of the controller, the most common approach is to introduce an objective function and optimize the PID parameters accordingly. This traditional approach is, however, not directly applicable because of the unknown plant model, and in particular the unknown varying gradient. A solution to this problem is still iterative learning. An objective function, which is accessible, is chosen as the first step for PID parameter optimization. Since the goal is to minimize the objective function, the control inputs will be updated along the steepest descending direction, namely the gradient, of the objective function. In other words, the PID parameters are chosen to directly reduce the objective function, and the objective function is treated as the plant output and used to update the PID parameters. When the gradient is varying and unknown, extra learning trials will be conducted to search for the best descending direction.

The chapter is organized as follows. Section 9.2 gives the formulation of the PID auto-tuning problem. Section 9.3 introduces the iterative learning tuning approach. Section 9.4 shows the comparative studies on benchmark examples. Furthermore,

Sect. 9.5 addresses the real-time implementation on a laboratory pilot plant. Section 9.6 presents the concludes on the work.

9.2 Formulation of PID Auto-tuning Problem

In this section we show that PID tuning can be formulated as an optimization problem.

9.2.1 PID Auto-tuning

In this work, we consider a fundamental PID controller in continuous or discrete time

$$C(s) = k_p + k_i \frac{1}{s} + k_d s$$

$$C(z) = k_p + k_i \frac{T_s z}{z-1} + k_d \frac{z-1}{T_s z},$$

where k_p is the proportional gain, k_i the integral gain, k_d is the derivative gain, s is the Laplace operator, z is the Z operator, T_s is the sampling period. Denote $\mathbf{k} = [k_p, k_i, k_d]^T$.

The aim of PID tuning is to find appropriate values for PID parameters such that the closed-loop response can be significantly improved when comparing with the open-loop response. Since the PID control performance is determined by PID parameters $\mathbf{k}$, by choosing a set of performance indices $\mathbf{x}$, for instance overshoot and settling time, there exists a relationship or mapping $\mathbf{f}$ between $\mathbf{x}$ and $\mathbf{k}$

$$\mathbf{x} = \mathbf{f}(\mathbf{k}).$$

The PID auto-tuning problem can be mathematically formulated to look for a set of $\mathbf{k}$ such that $\mathbf{x}$ meet the control requirements specified by $\mathbf{x}_r$. If the inverse mapping is available, we have $\mathbf{k} = \mathbf{f}^{-1}(\mathbf{x}_r)$. The mapping $\mathbf{f}$, however, is a vector-valued function of $\mathbf{x}$, $\mathbf{k}$ and the plant, and is in general a highly non-linear mapping. Thus, its inverse mapping in general is not analytically solvable. Above all, the most difficult problem in PID auto-tuning is the lack of a plant model, hence the mapping $\mathbf{f}$ is unknown or partially unknown.

The importance of PID and challenge in PID auto-tuning attracted numerous researchers who developed various auto-tuning or detuning methods, each with unique advantages and limitations. In this work we propose a new auto-tuning method using iterative learning and optimization, which complements existing PID tuning methods.

9.2.2 Performance Requirements and Objective Functions

An objective function, or cost function, quantifies the effectiveness of a given controller in terms of the closed-loop response, either in the time domain or frequency domain. A widely used objective function in PID auto-tuning is the integrated square error (ISE) function

$$J(\mathbf{k}) = \frac{1}{T - t_0} \int_{t_0}^{T} e^2(t, \mathbf{k}) dt, \tag{9.1}$$

where the error $e(t, \mathbf{k}) = r(t) - y(t, \mathbf{k})$ is the difference between the reference, $r(t)$, and the output signal of the closed-loop system, $y(t)$. T and t_0, with $0 \leq t_0 < T < \infty$, are two design parameters. In several auto-tuning methods such as IFT and ES, t_0 is set approximately at the time when the step response of the closed-loop system reaches the first peak. Hence, the cost function effectively places zero weighting on the initial transient portion of the response and the controller is tuned to minimize the error beyond the peak time. Similar objective functions, such as integrated absolute error, integrated time-weighted absolute error, integrated time-weighted square error,

$$\frac{1}{T - t_0} \int_{t_0}^{T} |e| dt, \quad \frac{1}{T - t_0} \int_{t_0}^{T} t |e| dt, \quad \frac{1}{T - t_0} \int_{t_0}^{T} t e^2 dt,$$

have also been widely used in the process of PID parameter tuning.

In many control applications, however, the transient performance, such as overshoot M_p, settling time t_s, rise time t_r, could be the main concern. An objective function that can capture the transient response directly, is highly desirable. For this purpose, we propose a quadratic function

$$J = (\mathbf{x}_r - \mathbf{x})^T Q(\mathbf{x}_r - \mathbf{x}) + \mathbf{k}^T R \mathbf{k}, \tag{9.2}$$

where $\mathbf{x} = [100 M_p, t_s, t_r]^T$, $\mathbf{x}_r$ is the desired $\mathbf{x}$, Q and R are two constant non-negative weighting matrices. The selection of Q and R matrices will yield effects on the optimization results, that is, produce different closed-loop responses. A larger Q highlights more on the transient performance, whereas a larger R gives more control penalty.

9.2.3 A Second-order Example

Consider a second order plant under a unity feedback with a PD controller. The PD controller and plant are respectively

$$C(s) = k_p + k_d s, \qquad G(s) = \frac{k}{s^2 + as + b}, \tag{9.3}$$

where k_p, k_d, k, a, b are all non-negative constants. The closed-loop system is stable if $a + d > 0$ and $b + p > 0$, where $p = kk_p$ and $d = kk_d$.

The non-linear mapping $\mathbf{f}$ between the closed-loop transient response $\mathbf{x} = [M_p, t_s]$ and the PD parameters $\mathbf{k} = [k_p, k_d]$ is derived in Section 9.7 Appendix, and shown in Fig. 9.1 and Fig. 9.2, where $a = 0.1$, $b = 0$, and $k = 1$. It can be seen that the

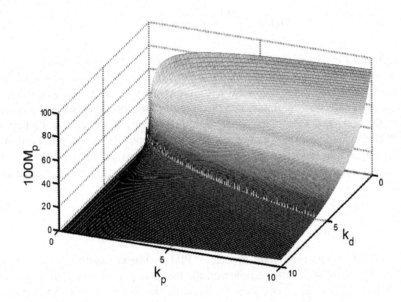

Fig. 9.1 The non-linear mapping between the peak overshoot $100M_p$ and PD gains (k_p, k_d) in continuous time

mapping $\mathbf{f}$ is non-linear or even discontinuous.

The non-linear mapping $\mathbf{f}$ for discrete-time control system can also be derived but is omitted here due to its complexity. Discretizing the plant in (9.3) with a sampling time $T_s = 0.1$ s, the non-linear mapping between M_p and (k_p, k_d) is shown in Fig. 9.3. It can be seen that there exist local minima in the surface, and the gradient may vary and take either positive or negative values.

On the other hand, it can also been seen from these figures that the transient responses may vary drastically while the control parameters only vary slightly. This indicates the importance for PID parameter auto-tuning and the necessity to find an effective tuning method.

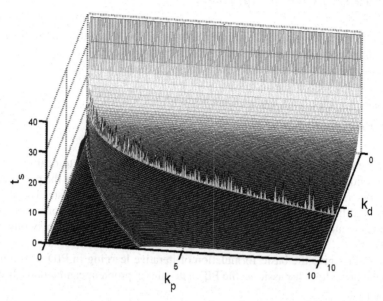

Fig. 9.2 The non-linear mapping between the settling time t_s and PD gains (k_p, k_d) in continuous time

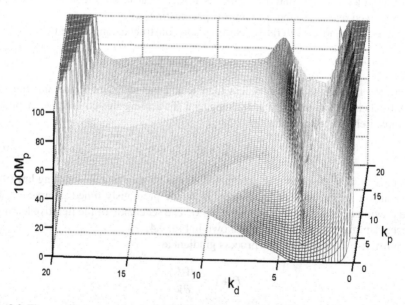

Fig. 9.3 The non-linear mapping between the peak overshoot $100M_p$ and PD gains (k_p, k_d) in discrete time. An upper bound of 100 is applied to crop the vertical values

9.3 Iterative Learning Approach

Iterative learning is adopted to provide a solution to the PID auto-tuning. The iterative learning offers the desirable feature that it can achieve the learning convergence even if the plant model is partially unknown.

9.3.1 Principal Idea of Iterative Learning

The concept of iterative learning was first introduced in control to deal with a repeated control task without requiring the perfect knowledge such as the plant model or parameters [4]. ILC learns to generate a control action directly instead of doing a model identification. The iterative learning mechanism updates the present control action using information obtained from previous control actions and previous error signals.

Let us first give the basic formulation of iterative learning in PID auto-tuning. From the preceding discussions, the PID auto-tuning problem can be described by the mapping

$$\mathbf{x} = \mathbf{f}(\mathbf{k}),$$

where $\mathbf{k} \in \Omega_k \subset \mathscr{R}^n$ and $\mathbf{x} \in \Omega_x \subset \mathscr{R}^m$, where n and m are integer numbers. The learning objective is to find a suitable set $\mathbf{k}$ such that the transient response $\mathbf{x}$ can reach a given region around the control specifications $\mathbf{x}_r$.

The principal idea of iterative learning is to construct a contractive mapping $\mathscr{A}$

$$\mathbf{x}_r - \mathbf{x}_{i+1} = A(\mathbf{x}_r - \mathbf{x}_i),$$

where the norm of A is strictly less than 1, and the subscript i indicates that the quantity is in the ith iteration or learning trial. To achieve this contractive mapping, a simple iterative learning law is

$$\mathbf{k}_{i+1} = \mathbf{k}_i + \Gamma_i(\mathbf{x}_r - \mathbf{x}_i), \qquad (9.4)$$

where $\Gamma_i \in \mathscr{R}^{n \times m}$ is a learning gain matrix. It can be seen that the learning law (9.4) generates a set of updated parameters from the previously tuned parameters, $\mathbf{k}_i$, and previous performance deviations $\mathbf{x}_r - \mathbf{x}_i$. The schematic of the iterative learning mechanism for PID auto-tuning is shown in Fig. 9.4.

When $n = m$, we define the process gradient as

$$F(\mathbf{k}) = \frac{\partial \mathbf{f}(\mathbf{k})}{\partial \mathbf{k}},$$

we can derive the condition for the contractive mapping $\mathscr{A}$

$$\mathbf{x}_r - \mathbf{x}_{i+1} = \mathbf{x}_r - \mathbf{x}_i - (\mathbf{x}_{i+1} - \mathbf{x}_i)$$

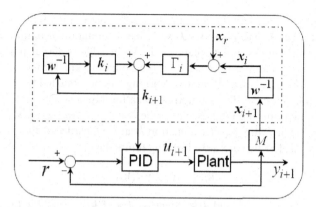

Fig. 9.4 The schematic block diagram of the iterative learning mechanism and PID control loop. The parameter correction is generated by the performance deviations $\mathbf{x}_r - \mathbf{x}_i$ multiplied by a learning gain Γ_i. The operator $\mathbf{z}^{-1}$ denotes one iteration delay. The new PID parameters $\mathbf{k}_{i+1}$ consist of the previous $\mathbf{k}_i$ and the correction term, analogous to a discrete-time integrator. The iterative learning tuning mechanism is shown by the block enclosed by the dashed line. r is the desired output and the block M is a feature extraction mechanism that records the required transient quantities such as overshoot from the output response y_{i+1}

$$= \mathbf{x}_r - \mathbf{x}_i - \frac{\partial \mathbf{f}(\mathbf{k}_i^*)}{\partial \mathbf{k}}(\mathbf{k}_{i+1} - \mathbf{k}_i)$$

$$= [I - F(\mathbf{k}_i^*)\Gamma_i](\mathbf{x}_r - \mathbf{x}_i), \qquad (9.5)$$

where $\mathbf{k}_i^* \in [\min\{\mathbf{k}_i, \mathbf{k}_{i+1}\}, \ \max\{\mathbf{k}_i, \mathbf{k}_{i+1}\}] \subset \Omega_k$. Therefore, we have a contractive mapping $\mathscr{A}$

$$\mathbf{x}_r - \mathbf{x}_{i+1} = A(\mathbf{x}_r - \mathbf{x}_i)$$

as far as the magnitude

$$|A| = |I - F(\mathbf{k}_i^*)\Gamma_i| \le \rho < 1. \qquad (9.6)$$

When $n > m$, there exists an infinite number of solutions because of redundancy in control parameters. With the extra degrees of freedom in PID, optimality can be exploited, for instance the shortest settling time, minimum peak overshoot, minimum values of control parameters, *etc*. A suitable objective function to be minimized could be a non-negative function $J(\mathbf{x}, \mathbf{k}) \ge 0$, where J is accessible, such as the quadratic one (9.2). The minimization is in fact a searching task

$$\min_{\mathbf{k} \in \Omega_k} J(\mathbf{x}, \mathbf{k}),$$

where $\mathbf{x} = \mathbf{f}(\mathbf{k})$. The optimal parameters $\mathbf{k}_{opt}$ can be obtained by differentiating J and computing

$$\frac{dJ}{d\mathbf{k}} = \frac{\partial J}{\partial \mathbf{x}}\frac{d\mathbf{x}}{d\mathbf{k}} + \frac{\partial J}{\partial \mathbf{k}} = \frac{\partial J}{\partial \mathbf{x}}F + \frac{\partial J}{\partial \mathbf{k}} = 0. \qquad (9.7)$$

In most cases $\mathbf{k}_{opt}$ can only be found numerically due to the highly non-linear relationship between $\mathbf{k}$ and quantities J, $\mathbf{x}$. A major limitation of optimization methods is the demand for complete plant-model knowledge or the mapping $\mathbf{f}$. On the contrary, iterative learning only requires the bounding knowledge of the process gradient. The principal idea of iterative learning can be extended to solving the optimization problem under the assumption that all gradient components with respect to $\mathbf{k}$ have known limiting bounds. Since the learning objective now is to directly reduce the value of J, the objective function J can be regarded as the process output to be minimized. The new iterative learning tuning law is

$$\mathbf{k}_{i+1} = \mathbf{k}_i - \gamma_i J(\mathbf{k}_i), \tag{9.8}$$

where $\gamma_i = [\gamma_{1,i}, \cdots, \gamma_{n,i}]^T$ and $J(\mathbf{k}_i)$ denotes $J(\mathbf{k}_i, \mathbf{f}(\mathbf{k}_i))$. To show the contractive mapping, note that

$$J(\mathbf{k}_{i+1}) = J(\mathbf{k}_i) + [J(\mathbf{k}_{i+1}) - J(\mathbf{k}_i)]$$

$$= J(\mathbf{k}_i) + \left(\frac{dJ(\mathbf{k}_i^*)}{d\mathbf{k}}\right)^T (\mathbf{k}_{i+1} - \mathbf{k}_i),$$

where $\mathbf{k}_i^* \in \Omega_k$ is in a region specified by $\mathbf{k}_i$ and $\mathbf{k}_{i+1}$. By substitution of the ILT law (9.8), we have

$$J(\mathbf{k}_{i+1}) = \left[1 - \left(\frac{dJ(\mathbf{k}_i^*)}{d\mathbf{k}}\right)^T \gamma_i\right] J(\mathbf{k}_i). \tag{9.9}$$

The convergence property is determined by the learning gains γ_i and the gradient $dJ/d\mathbf{k}$.

9.3.2 Learning Gain Design Based on Gradient Information

To guarantee the contractive mapping (9.9), the magnitude relationship must satisfy $|1 - D_i^T \gamma_i| \leq \rho < 1$, where

$$D_i = \left(\frac{dJ(\mathbf{k}_i^*)}{d\mathbf{k}}\right)$$

is the gradient of the objective function. The selection of learning gain γ_i is highly related to the prior knowledge on the gradient D_i. Consider three scenarios and assume $D_i = (D_{1,i}, D_{2,i}, D_{3,i})$, that is, a PID controller is used.

When D_i is known a priori, we can choose $\gamma_{j,i} = D_{j,i}^{-1}/3$. Such a selection produces the fastest learning convergence speed, that is, convergence in one iteration because $\|1 - D_i^T \gamma_i\| = 0$.

When the bounding knowledge and the sign information of $D_{j,i}$ are available, the learning convergence can also be guaranteed. For instanc, assume $0 < \alpha_j \leq D_{j,i} \leq$

$\beta_j < \infty$ for $\mathbf{k}_i^* \in \Omega_k$, where α_j and β_j are, respectively, the lower and upper bounds of the gradient components $D_{j,i}$. In such circumstances, choosing $\gamma_{j,i} = 1/3\beta_j$, the upper bound of the magnitude is $\rho = 1 - \alpha_1/3\beta_1 - \alpha_2/3\beta_2 - \alpha_3/3\beta_3 < 1$.

The most difficult scenario is when bounding functions or signs of $D_{j,i}$ are unknown. In order to derive the iterative learning convergence, it can be seen from (9.9) that we do not need the exact knowledge about the mapping $\mathbf{f}$. It is adequate to know the bounding knowledge (amplitude and sign) of the gradient $D(\mathbf{k}_i^*)$ or $F(\mathbf{k}_i^*)$. Although an invariant gradient is assumed for most iterative learning problems, the elements in $D(\mathbf{k}_i^*)$ may change sign and take either positive or negative values. Without knowing the exact knowledge of the mapping $\mathbf{f}$, we may not be able to predict the varying signs in the gradient. Now, in the iterative learning law (9.8) the learning gains will have to change signs according to the gradient. Here, the question is how to change the signs of the learning gains when we do not know the signs of the gradient components, *i.e.* how to search the direction of the gradient, which obviously can only be done in a model-free manner if $\mathbf{f}$ is unknown.

A solution to the problem with unknown gradient is to conduct extra learning trials to determine the direction of gradient or the signs of the learning gains *directly*

$$\gamma_i = [\pm\gamma_1, \cdots, \pm\gamma_n]^T, \qquad (9.10)$$

where γ_i are positive constants. From the derivation (9.5), when learning gains are chosen appropriately, $|A| < 1$ and the learning error reduces. On the other hand, if learning gains are chosen inappropriately, then $|A| > 1$ and the error increases after this learning trial. Therefore, several learning trials are adequate for the ILT mechanism to determine the correct signs of the learning gains.

In general, when there are two gradient components $(D_{1,i}, D_{2,i})$, there are 4 sets of signs $\{1,1\}, \{1,-1\}, \{-1,1\}$, and $\{-1,-1\}$, corresponding to all possibles signs of the gradient $(D_{1,i}, D_{2,i})$. In such circumstances, at most 4 learning trials are sufficient to find the steepest descending of the four control directions, as shown in Fig. 9.5.

Similarly, if there are three free tuning parameters, there will be 8 sets of signs in the gradient $(D_{1,i}, D_{2,i}, D_{3,i})$, as shown in Fig. 9.6. In general, if there are n control tuning parameters, there will be n gradient components $D_{j,i}$. Since each $\gamma_{j,i}$ takes either a positive or negative sign, there will be 2^n combinations and at most 2^n learning trials are required.

In addition to the estimation of the gradient direction or learning direction, the magnitudes of learning gains γ_i should also be adjusted to satisfy the learning convergence condition $|1 - D_i^T \gamma_i| \leq \rho < 1$. Since the gradient D_i is a function of PID parameters $\mathbf{k}_i$, the magnitude of D_i varies at different iterations. When the magnituds of the gradient components are unknown, extra learning trials will be needed to search for suitable magnitudes of learning gains.

In this work, we adopt a self-adaption rule [111] that scales learning gains up and down by a factor $\zeta = 1.839$, that is, each component $\gamma_{j,i}$ will be adjusted to $\gamma_{j,i}\zeta$ and $\gamma_{j,i}/\zeta$. Extra learning trials will be performed with the scaled learning gains. It is

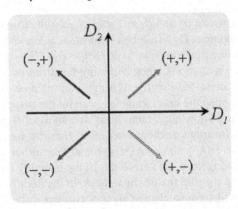

Fig. 9.5 There are four pairs of signs for the gradient (D_1, D_2) as indicated by the arrows. Hence there are four possible updating directions, in which one pair gives the fastest descending direction

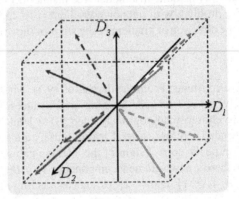

Fig. 9.6 There are three gradient components D_1, D_2 and D_3 with respect to three control parameters. Consequently, there are 8 possible tuning directions and at most 8 learning trials are required to find the correct updating direction

reported that the choice of such a scaling factor ζ will lead to a linear convergence rate [111, 110].

To facilitate searching and to reduce the number of trials, we can also estimate gradient components $D_{j,i}$ numerically using the values of the objective function and PID parameters obtained from the previous two iterations

$$\hat{D}_{j,i} = \frac{J(\mathbf{k}_{i-1}) - J(\mathbf{k}_{i-2})}{k_{j,i-1} - k_{j,i-2}}. \tag{9.11}$$

The learning gain can be revised accordingly as $\gamma_{j,i} = \lambda_j \hat{D}_{j,i}^{-1}$, where λ_j is a constant gain in the interval of $(0, 1]$. This method is in essence the Secant method along the iteration axis, which can effectively expedite the learning speed [153].

Since gradient direction is a critical issue in searching, and the approximation (9.11) may not always guarantee a correct sign, we can use the estimation result in (9.11) partially by retaining the magnitude estimation, while still searching the correct control direction

$$\gamma_{j,i} = \pm \lambda_j \left| \frac{k_{j,i-1} - k_{j,i-2}}{J(\mathbf{k}_{i-1}) - J(\mathbf{k}_{i-2})} \right|. \tag{9.12}$$

9.3.3 Iterative Searching Methods

Three iterative search methods are considered for ILT in this work with verifications and comparisons. They are M_0 – an exhaustive search method in directions and magnitude, M_1 – an exhaustive search method in directions, M_2 – a lazy search method.

M_0 does exhaustive searching in all 2^n directions and exhaustive searching in all 2^n magnitudes using self-adaptation with the factor ζ. Then, PID parameters will be updated using the set that generates the best closed-loop response or yields the biggest drop of J in that trial. With the best-tuned PID parameters as the initial setting, the ILT mechanism enters another run of exhaustive searching for the best response and best PID parameters. The searching process repeats until the stopping criterion is met.

This is the worst-case searching where neither the gradient directions nor the gradient magnitudes are available. For each run of searching the steepest descending direction, 4^n trials are performed. For PID tuning where $n = 3$, 64 trials are needed for one run of searching. Clearly, this search method is not efficient.

In M_1 the entire searching and updating process is similar to M_0 except that the magnitudes of learning gains are determined using the formula (9.12). Hence, it reduces the total number of trials from 4^n in M_0 to 2^n for each run of searching. For PID tuning where $n = 3$, only 8 trials are needed.

M_2 does exhaustive searching in directions in the first run of searching, and the magnitudes of learning gains are determined using the formula (9.12). Then, the steepest descending direction will be used for subsequent runs of searching, with the assumption that the mapping $\mathbf{f}$ is in general smooth and abrupt changes in the process gradient rarely occur. The exhaustive searching in directions will be activated again when the stopping criterion is met. The searching process will permanently stop if the stop criterion is still met after exhaustive searching in all directions.

The initial magnitudes of learning gains can be set as

$$[\gamma_{1,0}, \gamma_{2,0}, \gamma_{3,0}] = \frac{\gamma_0}{J_0}[k_{p,0}, k_{i,0}, k_{d,0}], \tag{9.13}$$

where γ_0 is a positive constant, and chosen to be 0.1 in this work. $k_{p,0}$, $k_{i,0}$, $k_{d,0}$ are initial values of PID parameters determined using any existing PID auto-tuning methods. J_0 is calculated with $k_{p,0}$, $k_{i,0}$, $k_{d,0}$ and the corresponding closed-loop response. Note that, by choosing $\gamma_0 = 0.1$, the adjustment size at each iteration is 10% of the initial magnitude $[k_{p,0}, k_{i,0}, k_{d,0}]/J_0$. A large γ_0 far above 0.1 may incur an unstable or rough learning process, and a small γ_0 far below 0.1 may lead to a slower learning rate.

9.4 Comparative Studies on Benchmark Examples

In this section we conduct comprehensive tests on 8 benchmark plant models. Four plant models G_1 to G_4 were used in [61] to compare and validate several iterative tuning methods

$$G_1(s) = \frac{1}{1+20s}e^{-5s},$$

$$G_2(s) = \frac{1}{1+20s}e^{-20s},$$

$$G_3(s) = \frac{1}{(1+10s)^8},$$

$$G_4(s) = \frac{1-5s}{(1+10s)(1+20s)},$$

where G_1 and G_2 have relatively large normalized time delay (NTD) which is defined as the ratio between the time delay and the time constant, G_3 has high-order repeated poles, and G_4 is the non-minimum phase. The other four plant models G_5 to G_8 were used in [43, 145] to validate their auto-tuning methods

$$G_5(s) = \frac{1}{(s+1)(s+5)^2}e^{-0.5s},$$

$$G_6(s) = \frac{1}{(25s^2+5s+1)(5s+1)}e^{-s},$$

$$G_7(s) = \frac{1}{(s^2+2s+3)(s+3)}e^{-0.3s},$$

$$G_8(s) = \frac{1}{(s^2+s+1)(s+2)^2}e^{-0.1s},$$

where G_5 is a high-order plant with medium NTD, G_6 is a high-order and moderately oscillatory plant with short NTD, G_7 is a high-order and heavily oscillatory plant with short NTD, and G_8 has both oscillatory and repeated poles.

To make fair comparisons, we choose initial learning gains with form (9.13) for PID parameters in all case studies. In all searching results, we use N to denote the total number of trials.

9.4.1 Comparisons Between Objective Functions

Objective function ISE (9.1) has been widely investigated and adopted in PID tuning. The quadratic objective function (9.2), on the other hand, has more weights to prioritize transient performance requirements. When control requirements are directly concerned with transient response such as M_p or t_s, we can only use the quadratic objective function (9.2). To show the effects of different objective functions and weight selections, we use plants $G_1 - G_4$. To make fair comparisons, the learning process starts from the same initial setting for PID gains generated by the Ziegler–Nichols (ZN) tuning method [9]. Exhaustive search method M_0 is employed. The tuning results through iterative learning are summarized in Table 9.1.

Table 9.1 Control performances of $G_1 - G_4$ using the proposed ILT method

Plant	J	PID Controller	$100M_p$	t_s	N
G_1	ISE	$3.59 + 0.13s^{-1} + 7.54s$	4.48	21.35	768
	$(100M_p)^2 + 0.5t_s^2$	$3.38 + 0.13s^{-1} + 7.05s$	1.99	12.17	1024
	$(100M_p)^2 + 0.1t_s^2$	$3.25 + 0.12s^{-1} + 6.30s$	0.63	12.83	832
	$(100M_p)^2 + 0.01t_s^2$	$3.71 + 0.11s^{-1} + 9.11s$	0.53	22.24	512
G_2	ISE	$0.93 + 0.031s^{-1} + 5.67s$	0.71	50.67	512
	$(100M_p)^2 + t_s$	$0.99 + 0.032s^{-1} + 6.87s$	1.06	47.99	1600
	$(100M_p)^2 + 0.2t_s$	$1.05 + 0.028s^{-1} + 9.79s$	0.29	82.74	512
	$(100M_p)^2 + 0.1t_s$	$1.03 + 0.029s^{-1} + 9.18s$	0.20	83.61	640
G_3	ISE	$0.64 + 0.012s^{-1} + 11.3s$	0.49	137.13	1024
	$(100M_p)^2 + t_s$	$0.76 + 0.013s^{-1} + 16.65s$	1.93	120.56	576
	$(100M_p)^2 + 0.2t_s$	$0.85 + 0.014s^{-1} + 25.77s$	0.66	212.04	192
	$(100M_p)^2 + 0.1t_s$	$0.83 + 0.014s^{-1} + 24.91s$	0.62	212.76	192
G_4	ISE	$5.01 + 0.092s^{-1} + 25.59s$	3.05	25.2	1216
	$(100M_p)^2 + 0.25t_s^2$	$4.31 + 0.075s^{-1} + 22.19s$	1.81	18.63	512
	$(100M_p)^2 + 0.1t_s^2$	$3.89 + 0.071s^{-1} + 22.28s$	1.70	20.56	384
	$(100M_p)^2 + 0.01t_s^2$	$4.51 + 0.075s^{-1} + 23.96s$	0.06	19.27	1216

For simplicity only M_p and t_s are taken into consideration in (9.2). The learning process stops when the decrement of an objective function between two consecutive iterations is smaller than $\varepsilon = 10^{-6}$ for (9.1) and $\varepsilon = 0.01$ for (9.2). In ISE, the parameters are $T = 100, 300, 500, 200$ s and $t_0 = 10, 50, 140, 30$ s, respectively [61]. Usually, the settling time is much greater than overshoot, thus $100M_p$ is used. When plants have a much bigger settling time, we can choose t_s instead of $(t_s)^2$ in the objective function, as shown in cases of G_2 and G_3. By scaling down t_s in objective functions, overshoot decreases as it is weighted more. Meanwhile, t_s increases as it is weighted less.

Comparing ISE and quadratic objective function, we can see that the latter offers more choices.

9.4.2 Comparisons Between ILT and Existing Iterative Tuning Methods

Now we compare iterative-learning-based tuning with other iterative tuning methods such as extremum seeking (ES) tuning and iterative feedback tuning (IFT). The ZN tuning method is also included for comparison. $G_1 - G_4$ and the same ISE [61] for each model are used. The PID controller parameters given by ZN are used as a starting point for ILT tuning. The exhaustive search method M_0 is employed in ILT.

Table 9.2 Control performances of $G_1 - G_4$ using methods ZN, IFT, ES and ILT

Plant	Method	PID controller	$100M_p$	t_s
G_1	ZN	$4.06 + 0.44s^{-1} + 9.38s$	46.50	47.90
	IFT	$3.67 + 0.13s^{-1} + 7.74s$	5.38	21.38
	ES	$3.58 + 0.13s^{-1} + 7.68s$	3.31	21.35
	ILT	$3.59 + 0.13s^{-1} + 7.54s$	4.48	21.35
G_2	ZN	$1.33 + 0.043s^{-1} + 10.30s$	21.75	109.4
	IFT	$0.93 + 0.031s^{-1} + 5.64s$	0.80	50.33
	ES	$1.01 + 0.032s^{-1} + 7.23s$	1.37	76.61
	ILT	$0.93 + 0.031s^{-1} + 5.67s$	0.71	50.67
G_3	ZN	$1.10 + 0.015s^{-1} + 20.91s$	13.95	336.90
	IFT	$0.66 + 0.012s^{-1} + 12.08s$	0.98	132.05
	ES	$0.68 + 0.013s^{-1} + 13.30s$	0.96	130.41
	ILT	$0.64 + 0.012s^{-1} + 11.30s$	0.49	137.13
G_4	ZN	$3.53 + 0.21s^{-1} + 14.80s$	53.70	86.12
	IFT	$3.03 + 0.065s^{-1} + 18.42s$	0.55	28.74
	ES	$3.35 + 0.068s^{-1} + 21.40s$	0.18	29.80
	ILT	$5.01 + 0.092s^{-1} + 25.59s$	3.05	25.21

The results verify that ILT, ES and IFT give very similar responses that are far superior than that of ZN tuning method. Figure 9.7 shows the ILT results for G_1, (a) shows the decreasing objective function J, (b) shows decreasing performance indices M_p and t_s, (c) shows the variations of the PID parameters, and (d) compares step responses with four tuning methods. Figure 9.8 shows the searching results of the gradient directions and the variations of the learning gains through self-adaptation. It can be seen that the gradients undergo changes in signs, hence it is in general a difficult and challenging optimization problem.

Table ?? summarizes the comparative results with all four plants $G_1 - G_4$ when 4 tuning methods were applied. The iteration numbers when applying ILT for $G_1 - G_4$ are 786, 512, 1024 and 1216, respectively.

Although ILT shares similar performance as ES and IFT, it is worth highlighting some important factors in the tuning process. In ES tuning, there are more than 10 design parameters to be set properly. From [61], design parameters take rather specific values. In IFT, the initial values of the PID parameters must be chosen in such a way as to give an initial response that is very slow and with no overshoot.

Further, in IFT the transient performance is purposely excluded from the objective function by choosing a sufficiently large t_0. On the contrary, in ILT only initial learning gains need to be preset, and we choose all three learning gains with a uniform value $\gamma_0 = 0.1$ in (9.13) for the ease of ILT design. In fact, by choosing initial learning gains with different values, we can achieve much better responses than those in Table ??. Needless to mention that ILT can handle both types of objective functions (9.1) and (9.2) with the flexibility to highlight transient behaviors.

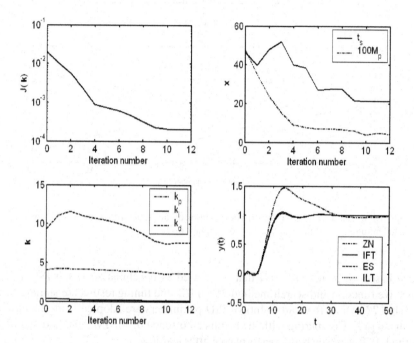

Fig. 9.7 ILT performance for G_1. (a) The evolution of the objective function; (b) The evolution of overshoot and settling time; (c) The evolution of PID parameters; (d) The comparisons of step responses among ZN, IFT, ES and ILT, where IFT, ES and ILT show almost the same responses

9.4.3 Comparisons Between ILT and Existing Auto-tuning Methods

PID auto-tuning methods [43, 145, 9] provided several effective ways to determine PID parameters. In this subsection we compare ILT with the auto-tuning method [9] based on internal model control (IMC), and the auto-tuning method [145] based on pole-placement (PPT) that shows superior performance to [43]. Comparisons are

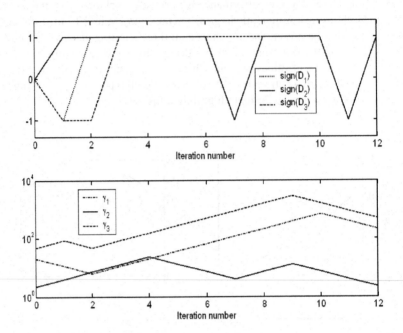

Fig. 9.8 ILT searching results for G_1. (a) The evolution of the gradient directions; (b) The evolution of the magnitudes of learning gains with self-adaptation

made based on plants $G_5 - G_8$ that were used as benchmarks. Using ISE as the objective function and search method M_0 in ILT, the tuning results are summarized in Table **??**, where the start points of PID parameters are adopted from the tuning results in [43]. Comparing with the results auto-tuned by the IMC method and PPT method, ILT achieves better performance after learning.

9.4.4 Comparisons Between Search Methods

Now we investigate the effects of search methods M_0, M_1 and M_2. Plants $G_1 - G_8$ are used. The objective function to be used is

$$J = (100M_p)^2 + \frac{t_s^2}{100}. \tag{9.14}$$

Set $\lambda_j = 0.2$, $j = 1, 2, 3$ in (9.12) for M_1 and M_2. The stopping criterion for ILT is $\varepsilon = 0.01$. The control performance after learning is summarized in Tables 9.4 and 9.5.

Table 9.3 Control performances of $G_5 - G_8$ using IMC, PPT and ILT methods

Plant	Method	PID Controller	$100M_p$	t_s
G_5	IMC	$33.46 + 16.67s^{-1} + 8.54s$	2.09	5.71
	PPT	$27.02 + 21.14s^{-1} + 6.45s$	12.12	5.01
	ILT	$31.96 + 18.81s^{-1} + 11.01s$	1.31	2.09
G_6	IMC	$0.88 + 0.066s^{-1} + 1.20s$	3.46	74.61
	PPT	$0.44 + 0.074s^{-1} + 2.54s$	5.10	49.81
	ILT	$0.57 + 0.075s^{-1} + 2.78s$	1.17	23.91
G_7	IMC	$7.06 + 5.29s^{-1} + 1.64s$	5.15	8.37
	PPT	$3.89 + 5.39s^{-1} + 2.15s$	3.21	5.32
	ILT	$4.51 + 5.40s^{-1} + 2.53s$	0.44	2.93
G_8	IMC	$2.79 + 1.33s^{-1} + 1.19s$	3.01	14.74
	PPT	$1.50 + 1.37s^{-1} + 1.72s$	3.04	9.44
	ILT	$2.18 + 1.48s^{-1} + 2.34s$	1.18	4.61

Table 9.4 Control performance of $G_1 - G_8$ using search methods M_0, M_1, M_2

Plant	M_0			M_1			M_2		
	$100M_p$	t_s	N	$100M_p$	t_s	N	$100M_p$	t_s	N
G_1	0.53	22.24	512	0.24	22.13	160	0.23	22.23	29
G_2	1.80	78.64	256	0.73	79.93	72	1.47	79.91	43
G_3	1.81	121.29	192	1.36	209.77	48	1.81	205.35	41
G_4	0.06	19.27	1216	0	15.92	120	0	22.93	41
G_5	0.282	2.16	640	0.21	2.44	48	0	3.34	70
G_6	0.605	25.04	320	0.03	48.04	40	0.03	48.04	19
G_7	0	6.05	192	0	6.13	32	0	6.15	28
G_8	0.19	8.25	768	0	10.53	32	0	10.74	18

Table 9.5 Final controllers for $G_1 - G_8$ by using search methods M_0, M_1, M_2

Plant	M_0	M_1	M_2
G_1	$3.59 + \frac{0.13}{s} + 7.54s$	$3.42 + \frac{0.11}{s} + 8.77s$	$3.38 + \frac{0.12}{s} + 7.80s$
G_2	$0.93 + \frac{0.029}{s} + 9.43s$	$0.94 + \frac{0.030}{s} + 7.28s$	$1.05 + \frac{0.031}{s} + 8.98s$
G_3	$0.75 + \frac{0.013}{s} + 16.26s$	$0.87 + \frac{0.012}{s} + 24.73s$	$0.72 + \frac{0.012}{s} + 21.42s$
G_4	$4.51 + \frac{0.075}{s} + 23.96s$	$5.99 + \frac{0.097}{s} + 30.70s$	$6.92 + \frac{0.10}{s} + 39.18s$
G_5	$31.52 + \frac{18.31}{s} + 10.76s$	$29.63 + \frac{17.65}{s} + 10.19s$	$17.00 + \frac{13.68}{s} + 0.54s$
G_6	$0.51 + \frac{0.070}{s} + 2.14s$	$0.60 + \frac{0.069}{s} + 2.17s$	$0.60 + \frac{0.069}{s} + 2.17s$
G_7	$5.34 + \frac{4.86}{s} + 1.28s$	$5.02 + \frac{4.81}{s} + 1.31s$	$4.11 + \frac{4.81}{s} + 1.31s$
G_8	$2.03 + \frac{1.27}{s} + 2.71s$	$2.13 + \frac{1.21}{s} + 0.94s$	$1.74 + \frac{1.21}{s} + 0.94s$

It can be seen that M_1 and M_2 achieve similar performance, which is slightly inferior to M_0. However, comparing with M_0 the learning trial numbers in M_1 and M_2 have been significantly reduced.

9.4.5 ILT for Sampled-data Systems

A promising feature of ILT is the applicability to sampled-data or discrete-time systems. To illustrate how ILT works for digital systems, consider plant G_4 that can be discretized using a sampler and zero-order hold

$$G_4(z) = \frac{\left(-2.5e^{-.05T_s} + 1.5e^{-.1T_s} + 1\right)z + e^{-.15T_s} - 2.5e^{-0.1T_s} + 1.5e^{-.05T_s}}{z^2 - (e^{-0.1T_s} + e^{-0.05T_s})z + e^{-0.15T_s}},$$

where the sampling period $T_s = 0.1$ s. The digital PID controller is used. Choose again (9.14) as the objective function, and use ZN to generate the initial values for PID parameters. The closed-loop responses using ILT are summarized in Table ??. For comparison all three search methods M_0, M_1 and M_2 are used. It can be seen that in all cases the control responses have been improved drastically, especially the reduction in overshoot.

Table 9.6 Digital control results. Initial performance is achieved by ZN-tuned PID. Final performance is achieved by ILT

T_s	Initial performance		Method	Final	Final performance		
	$100M_p$	t_s		k_p, k_i, k_d	$100M_p$	t_s	N
0.01	39.16	64.14	M_0	3.56, 0.10, 21.62	0.13	14.77	896
			M_1	4.04, 0.12, 25.26	0.43	11.25	80
			M_2	3.24, 0.11, 21.12	1.12	17.64	43
0.05	39.52	63.90	M_0	3.45, 0.10, 20.89	0.057	15.45	768
			M_1	4.04, 0.12, 25.26	0.55	10.85	80
			M_2	3.25, 0.11, 21.12	1.10	17.35	43
0.2	40.92	63.00	M_0	3.53, 0.11, 21.76	0.060	13.60	1024
			M_1	4.08, 0.15, 28.21	1.02	11.80	96
			M_2	2.80, 0.090, 16.41	0.70	20.20	56
0.5	44.26	77.00	M_0	3.39, 0.11, 21.60	0.00	11.50	832
			M_1	3.48, 0.13, 24.05	0.97	15.50	72
			M_2	2.69, 0.086, 15.61	0.74	20.00	48
2	76.96	88.00	M_0	2.22, 0.082, 17.60	1.08	34.00	576
			M_1	2.19, 0.078, 15.22	0.32	34.00	128
			M_2	2.16, 0.075, 13.94	0.0020	18.00	81

9.5 Real-time Implementation

In order to show the applicability and effectiveness of the proposed ILT, real-time experiment has been carried out on a coupled tank.

9.5.1 Experimental Setup and Plant Modelling

The coupled-tank equipment consists of two small perspex tower-type tanks and interconnected through a hole that yields a hydraulic resistance (Fig. 9.9). The experimental setup in this work was configured such that the level is measured from tank 2 while the water is pumped into tank 1 as the control input. The outlet of tank 2 is used to discharge the water into the reservoir. The measurement data is collected from the coupled tank using an NI data acquisition card USB-6008. The control method is programmed using Labview. A window-type smoothing filter using 100 samples is implemented to mitigate measurement noise. The sampling period is 0.125 s.

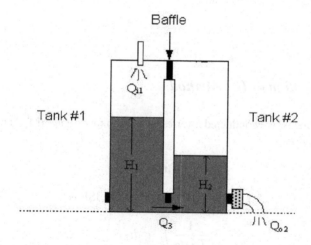

Fig. 9.9 Diagram of coupled-tank apparatus

A step response is conducted to approximate the coupled-tank dynamics with a first order plus time delay model [9]

$$G(s) = \frac{k}{\tau s + 1} e^{-sL}, \tag{9.15}$$

where k is the plant DC gain, τ is the time constant, L is the transportation lag. As shown in Fig. 9.10, after conducting a step response the obtained plant is

$$G(s) = \frac{168}{1+65s}e^{-3.6s}. \tag{9.16}$$

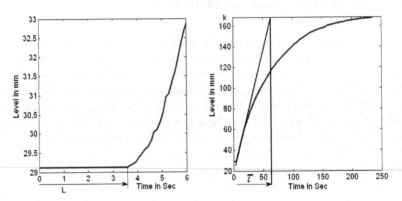

Fig. 9.10 Step-response-based modelling

9.5.2 Application of ILT Method

In the experiments, we conducted a series of tests to investigate ILT. The quadratic objective function is

$$J = (100M_p)^2 + qt_s^2,$$

where two values $q = 0.1$ and $q = 0.01$ are used. The ISE is

$$\frac{1}{T-t_0}\int_{t_0}^{T} e^2 dt,$$

where $T = 124$ s and $t_0 = 27.4$ s. All three search methods M_0, M_1 and M_2 are applied.

It is observed in the experiment that the calculated gradient components (9.11) or (9.12) may be singular sometimes. This is due to the presence of measurement noise and the closeness of the values of J at two adjacent iterations. On the other hand, the learning process may become sluggish when the PID parameters at two adjacent iterations are too close, yielding very much lower learning gains $\gamma_{j,i}$. To solve these two problems, a constraint is applied when updating the learning gains

$$c_1|k_{j,i}| \leq \gamma_{j,i} \leq c_2|k_{j,i}|, \tag{9.17}$$

where $0 \leq c_1 < c_2 \leq 1$. The upper and lower learning gain bounds are made in proportion to PID parameters. The rationale is clear. When a controller parameter is bigger, we can update it with a larger bound without incurring drastic changes. If setting absolute bounds for the learning gain, an overly small bound would limit the parameter updating speed, and an overly large bound would make the constraint ineffective. In the real-time application we consider 2 sets of boundaries

C_1: $c_1 = 0.02$ $c_2 = 0.2$,
C_2: $c_1 = 0.05$ $c_2 = 0.4$.

The initial PID controller is found using the ZN tuning method

$$0.11 + \frac{0.011}{s} + 0.26s.$$

The transient performance with ZN-tuned PID is $100M_p = 30.02$ and $t_s = 58.73$ s. ILT is applied to improve the performance.

9.5.3 Experimental Results

The results are summarized in Table 9.7 where the first column denotes the methods used. It can be seen that search methods M_1 and M_2 can significantly reduce the number of learning iterations, while M_p and t_s can be maintained at almost the same level. By changing the weight for t_s, the final overshoot M_p and settling time t_s can be adjusted. The constraint C_2 can also reduce the trial number, because higher limits in learning gains will expedite the learning progress. In experiments, it is observed that M_2 can be further simplified by removing the last run of exhaustive searching, so long as the variation of J reaches the preset threshold ε.

9.6 Conclusion

A new PID auto-tuning method is developed and compared with several well-established auto-tuning methods including ZN, IFT, ES, PPT, IMC. Iterative learning tuning provides a sustained improvement in closed-loop control performance, and offers extra degrees of freedom in specifying the transient control requirements through a new objective function.

The iterative learning tuning method proposed in this work can be further extended to various optimal designs for controllers, owing to its model-free nature. For instance, by iterative searching and task repeatability, we can tune the parameters of lead and lag compensators, filters, observers, and use both time-domain and frequency-domain performance indices in objective functions.

The future issues associated with the iterative-learning-tuning approach we are looking into include the robustness of ILT when the control process is not completely repeatable, the incorporation of various practical constraints, improvement of searching efficiency for high-dimension multi-loop control systems, as well as ILT for closed-loop systems with lower stability margins such as the sampled-data system with large time delays.

Table 9.7 Experimental results

Met	Constraint	J	PID controller	$100M_p$	t_s	N
M_0	C_1	$(100M_p)^2 + 0.01t_s^2$	$0.11 + 0.0029s^{-1} + 0.26s$	0.089	16.65	384
	C_1	ISE	$0.16 + 0.0045s^{-1} + 0.32s$	4.18	34.04	56
	C_2	ISE	$0.14 + 0.0039s^{-1} + 0.20s$	4.29	30.22	32
M_1	C_1	$(100M_p)^2 + + 0.01t_s^2$	$0.21 + 0.0036s^{-1} + 0.43s$	0.11	16.58	56
	C_2	$(100M_p)^2 + 0.01t_s^2$	$0.16 + 0.0041s^{-1} + 0.34s$	0.78	10.45	24
	C_1	$(100M_p)^2 + 0.1t_s^2$	$0.26 + 0.0037s^{-1} + 0.45s$	1.40	9.35	48
	C_2	$(100M_p)^2 + 0.1t_s^2$	$0.22 + 0.0036s^{-1} + 0.44s$	1.44	10.10	32
	C_1	ISE	$0.26 + 0.0040s^{-1} + 0.62s$	1.30	10.50	15
	C_2	ISE	$0.23 + 0.0042s^{-1} + 0.56s$	2.19	12.98	13
M_2	C_1	$(100M_p)^2 + 0.01t_s^2$	$0.32 + 0.0034s^{-1} + 0.76s$	0.27	17.30	17
	C_2	$(100M_p)^2 + 0.01t_s^2$	$0.30 + 0.0034s^{-1} + 0.72s$	0.50	16.63	14
	C_1	$(100M_p)^2 + 0.1t_s^2$	$0.30 + 0.0038s^{-1} + 0.72s$	1.30	16.33	14
	C_2	$(100M_p)^2 + 0.1t_s^2$	$0.23 + 0.0037s^{-1} + 0.56s$	1.71	15.99	10

9.7 Appendix

The non-linear mapping **f** will be explored for three cases with underdamped, overdamped and critical-damped characteristics.

9.7.1 Underdamped Case

When the system parameters satisfy $(a + kk_d)^2 - 4(b + kk_p) < 0$, the closed-loop system shows underdamped response with two complex poles. The system response in the time domain is

$$y(t) = A - A\sqrt{1 + \lambda^2}e^{-\alpha t}\cos(\beta t - \tan^{-1}\lambda), \qquad (9.18)$$

where

$$A = \frac{p}{b+p}, \quad \alpha = \frac{a+d}{2},$$

$$\beta = \frac{1}{2}\sqrt{4(b+p)-(a+d)^2},$$

$$\lambda = \frac{ap-2bd-pd}{p\sqrt{4(b+p)-(a+d)^2}}$$

with $p = kk_p$ and $d = kk_d$. It is easy to see that the steady-state output is A and

$$M_p = -\sqrt{1+\lambda^2}e^{-\alpha t_p}\cos\left(\beta t_p - \tan^{-1}\lambda\right), \qquad (9.19)$$

$$t_s = \frac{1}{\alpha}\ln\left(50\sqrt{1+\lambda^2}\right), \qquad (9.20)$$

where the peak time is

$$t_p = \begin{cases} \frac{1}{\beta}\tan^{-1}\frac{\beta\lambda-\alpha}{\beta+\alpha\lambda} & \text{if } \frac{\beta\lambda-\alpha}{\beta+\alpha\lambda} \geq 0, \\ \frac{1}{\beta}\left[\pi+\tan^{-1}\frac{\beta\lambda-\alpha}{\beta+\alpha\lambda}\right] & \text{if } \frac{\beta\lambda-\alpha}{\beta+\alpha\lambda} < 0. \end{cases}$$

Note that the settling time t_s, corresponding to 2% of the final steady-state value, is calculated from the worst-case condition

$$\left|\sqrt{1+\lambda^2}e^{-\alpha t}\cos\left(\beta t - \tan^{-1}\lambda\right)\right| < 0.02$$

by ignoring the factor $\cos\left(\beta t - \tan^{-1}\lambda\right)$.

9.7.2 Overdamped Case

When the system parameters satisfy $(a+kk_d)^2 - 4(b+kk_p) > 0$, the closed-loop shows overdamped response and the closed-loop system has two distinct real poles. The system response in the time domain is

$$y(t) = A - \frac{1}{2}\left[\frac{d}{\beta} - \frac{p}{\beta(\alpha+\beta)}\right]e^{-(\alpha+\beta)t} + \frac{1}{2}\left[\frac{d}{\beta} - \frac{p}{\beta(\alpha-\beta)}\right]e^{-(\alpha-\beta)t}, (9.21)$$

where

$$\alpha = \frac{a+d}{2}, \quad \beta = \frac{1}{2}\sqrt{(a+d)^2 - 4(b+p)}.$$

The relationship $\alpha > \beta$ always holds since $b+p > 0$. Therefore the steady state is still A.

First, derive the peak overshoot. From

$$\frac{dy}{dt} = \frac{1}{2\beta}[(d\alpha+d\beta-p)e^{-(\alpha+\beta)t} + (-d\alpha+d\beta+p)e^{-(\alpha-\beta)t}],$$

and let $dy/dt = 0$ we can obtain the peak time

$$t_p = -\frac{1}{2\beta}\ln(\frac{d\alpha - d\beta - p}{d\alpha + d\beta - p}).$$

The output has overshoot if and only if a, b, p, d lie in the domain

$$\mathscr{D} = \{(a, b, p, d) \mid d\alpha - d\beta - p > 0\}.$$

When the system parameters lie in the domain $\mathscr{D}$,

$$M_p = \frac{y(t_p) - A}{A} = \frac{1}{2\beta p}\left\{ \left(\frac{d\alpha - d\beta - p}{d\alpha + d\beta - p}\right)^{\frac{\alpha - \beta}{2\beta}} (\alpha + \beta)(d\alpha - d\beta - p) \right.$$
$$\left. - \left(\frac{d\alpha - d\beta - p}{d\alpha + d\beta - p}\right)^{\frac{\alpha + \beta}{2\beta}} (\alpha - \beta)(d\alpha + d\beta - p) \right\}.$$

Now derive the settling time. Let $|\frac{y(t)-A}{A}| < 0.02$ for all $t \geq t_s$. If the overshoot exceeds 2%, then t_s is the bigger real root of the following equation

$$-\frac{1}{2}\left[\frac{d}{\beta} - \frac{p}{\beta(\alpha + \beta)}\right] e^{-(\alpha + \beta)t} + \frac{1}{2}\left[\frac{d}{\beta} - \frac{p}{\beta(\alpha - \beta)}\right] e^{-(\alpha - \beta)t} = 0.02\frac{p}{b + p};$$

otherwise, t_s is the unique real root of the following equation

$$-\frac{1}{2}\left[\frac{d}{\beta} - \frac{p}{\beta(\alpha + \beta)}\right] e^{-(\alpha + \beta)t} + \frac{1}{2}\left[\frac{d}{\beta} - \frac{p}{\beta(\alpha - \beta)}\right] e^{-(\alpha - \beta)t} = -0.02\frac{p}{b + p}.$$

Note that in neither cases can we obtain the analytic expression of t_s.

9.7.3 Critical-damped Case

When the system parameters satisfy $(a + kk_d)^2 - 4(b + kk_p) = 0$, the closed-loop system shows a critical-damped response with two identical real poles. The system response in the time domain is

$$y(t) = A - \frac{p + (p - d\alpha)\alpha t}{b + p}e^{-\alpha t}, \tag{9.22}$$

with the steady-state A.

First, derive the peak overshoot. Letting

$$\frac{dy}{dt} = \frac{\alpha^2(d + (p - d\alpha)t)}{b + p}e^{-\alpha t} = 0$$

yields a unique solution

$$t_p = \frac{-d}{p - d\alpha} = \frac{-2d}{2p - ad - d^2},$$

implying that the overshoot exists if and only if $a^2 < 4b + d^2$. The overshoot is

$$M_p = \frac{d^2(a+d)^2 - p(-a^2 + 4b + 3d^2 + 2ad)}{p(-a^2 + 4b + d^2)} e^{\frac{-2d(a+d)}{-a^2 + 4b + d^2}}.$$

Next, derive the settling time. If the overshoot exceeds 2%, then t_s is the bigger real root of the following equation

$$-\frac{p + (p - d\alpha)\alpha t}{b + p} e^{-\alpha t} = 0.02 \frac{p}{b + p};$$

otherwise, t_s is the unique real root of the following equation

$$-\frac{p + (p - d\alpha)\alpha t}{b + p} e^{-\alpha t} = -0.02 \frac{p}{b + p}.$$

Note that in neither cases can we obtain the analytic expression of t_s.

Chapter 10
Calibration of Micro-robot Inverse Kinematics Using Iterative Learning Approach

Abstract In this chapter, an iterative learning identification method is developed to estimate the constant parameters of micro-robots with highly non-linear inverse kinematics. Based on the iterative learning theory, the new identification mechanism is able to approximate the actual parameters iteratively using previous estimates and corresponding kinematics mapping. A novel property of the new identification method is that, by using only one precise calibration sample, it is possible to estimate all unknown parameters non-linear in the inverse kinematics. A multi-link closed-chain robotic manipulator is utilized as an example to verify this method. Sensitivity analysis with respect to the calibration precision is also conducted.

10.1 Introduction

The calibration of various models of robots has been investigated extensively. In [163], a complete and parametrically continuous (CPC) error model has been derived for robot calibration. In [57], a survey was conducted in the field of calibration and accuracy of kinematic and dynamic models for manipulation robots. In [101], experimental investigations have been conducted on a 6 degree of freedom (6-DOF) platform for precision 3D micromanipulation. In [48] a new method has been developed for identifying the principal screws of the spatial 3-DOF parallel manipulators. Techniques have been provided in [87] for the automatic identification of constraints in computer models of 3D assemblies. In [165], a method has been developed for finding and classifying all the singularities of an arbitrary non-redundant mechanism. In [31], an algorithm has been developed for the branch identification of a large variety of multi-loop linkages and manipulators. In [124], sensitivity from position and orientation coordinates of a manipulator gripper to errors of kinematic parameters has been analyzed.

Nowadays, the sizes of the robots become smaller and smaller. Many micro-robots must operate at the micrometer level. In [138] the advantages of micro-mechanical systems and analsis of the scaling of forces in the micro-domain have

169

been explored. In order to achieve high-precision position control in micro-robot manipulators, the accurate inverse kinematics model is indispensable. When a precision control task is to be performed at the micrometer level, it is necessary to have the inverse kinematics model accurate at least up to the same precision level or even higher, say preferably up to submicrometer or even nanometer level. Any tiny discrepancy may lead to the failure of the control task.

This imposes a technically challenging task: how to precisely identify calibration is a tedious, time-consuming and costly process for mass-produced precision equipment, especially at the micrometer or nanometer scale. It is highly desirable that such identification requires as few calibrated samples as possible, or even one single sample. Considering the variety and complexity in robotic inverse kinematics models, it is also of crucial importance to find a generic scheme that can equally handle all types of inverse kinematics or kinematics. Moreover, in most cases the parameters to be identified are non-linear in the robotic inverse kinematics. Obviously the standard least squares estimation does not apply. Non-linear least squares are also difficult to apply due to the various types of highly non-linear structural relationships in the parametric space, the demand for a large number of samples, and a large amount of computation. It is highly desirable to develop a powerful scheme that can accurately accomplish the inverse kinematics identification task with the minimum effort.

In this work we develop a new method – iterative learning identification, which is able to effectively and efficiently estimate the unknown parameters of the inverse kinematics. The underlying idea of iterative learning identification is motivated by, and hence analogous to, a well-established control methodology – iterative learning control [38, 51, 60, 85, 86, 90, 102, 130, 135, 150, 154, 155]. Let us briefly explain the basic idea of iterative learning identification. The actual inverse kinematics will be viewed as a reference model, and the calibration results will be recorded as the reference model output signals. The unknown parameters in the actual inverse kinematics, on the other hand, will be viewed as virtual control inputs that generate the outputs of the inverse kinematics. An estimation model is constructed according to the structure of the inverse kinematics and contains an equal number of adjustable parameters as the adjustable control inputs. An iterative learning identification mechanism updates the estimation model inputs "iteratively", such that the estimation model output approximates the calibration results. As far as the Jacobian matrix of the inverse kinematics is non-singular around the calibrated position, the convergence of the output implies the convergence of "inputs", that is, the convergence of the parameter estimates.

This chapter is organized as follows. Section 10.2 briefs on the simplest iterative learning control approach. Section 10.3 details the iterative learning identification with convergence analysis. Section 10.4 conducts the sensitivity analysis with regards to the iterative learning identification. Section 10.5 provides an illustrative example to verify the proposed method.

10.2 Basic Idea of Iterative Learning

The simplest formulation of iterative learning control is as follows. Consider a mapping from input $\mathbf{u}(t) \in R^m$ to the output $\mathbf{y}(t) \in R^n$

$$\mathbf{y}(t) = \mathbf{f}(\mathbf{u}(t), t) \quad \forall t \in [0, T],$$

where $\mathbf{f} \in R^n$. The control objective is to find a suitable input $\mathbf{u}$ (denoted as $\mathbf{u}_d$) such that $\mathbf{y}$ follows perfectly a given target trajectory $\mathbf{y}_r$, i.e.

$$\mathbf{y}_r = \mathbf{f}(\mathbf{u}_d, t).$$

It should be noted that, even if $\mathbf{f}$ is known a priori, one may not be able to find an analytic solution or the closed form of $\mathbf{u}$ as a function of $\mathbf{y}$.

Practically, $\mathbf{f}$ itself may have uncertainties and that makes it impossible to calculate $\mathbf{u}_d$ by inverting $\mathbf{f}^{-1}(\mathbf{y}_r, t)$. Iterative learning control theory suggests a simple iterative structure to effectively find the desired signal u_d

$$\mathbf{u}_{i+1} = \mathbf{u}_i + \Gamma(\mathbf{y}_r - \mathbf{y}_i),$$

where i denotes the iteration number, and $\Gamma \in R^{m \times n}$ is a learning gain matrix. The convergence condition of $\mathbf{u}_i$ to $\mathbf{u}_d$ is dependent on the Jacobian matrix

$$0 < \left\| I - \Gamma \frac{\partial \mathbf{f}}{\partial \mathbf{u}} \right\| \leq \rho < 1.$$

This condition can be easily met by choosing an appropriate gain matrix Γ when $n = m$, that is, the Jacobian matrix is square. When $n > m$, the Jacobian is left invertible, the least squares approach can be applied, as we will show later. If $n < m$, the Jacobian matrix is right invertible, the solution of $\mathbf{u}_i$ is not unique and other optimization criteria can be introduced.

10.3 Formulation of Iterative Identifications

Now let us see how we can apply the same idea to the identification problem of micro-robot inverse kinematics. Consider a multi-link closed-chain robotic manipulator (Fig. 10.1), its inverse kinematics model is given by

$$\begin{cases} l_1 = z_{C_{12}} - \sqrt{b^2 - \Theta(x^-_{C_{15}}) - \Theta(y^+_{C_{12}})} \\ l_2 = z_{C_{12}} - \sqrt{b^2 - \Theta(x^-_{C_{12}}) - \Theta(y^-_{C_{12}})} \\ l_3 = z_{C_{34}} - \sqrt{b^2 - x^2_{C_{34}} - (y_{C_{34}} - a)^2} \\ l_4 = z_{C_{34}} - \sqrt{b^2 - \Theta(x^+_{C_{34}}) - \Theta(y^-_{C_{34}})} \\ l_1 = z_{C_{46}} - \sqrt{b^2 - \Theta(x^+_{C_{30}}) - \Theta(y^+_{C_{56}})} \\ l_6 = z_{C_{56}} - \sqrt{b^2 - x^2_{C_{56}} - (y_{C_{56}} + a)^2}, \end{cases} \qquad (10.1)$$

where $\Theta(x^+_{C_{ij}}) = (x_{C_{ij}} + a\,\cos\frac{\pi}{6})^2$, $\Theta(x^-_{C_{ij}}) = (x_{C_{ij}} - a\,\cos\frac{\pi}{6})^2$, $\Theta(y^+_{C_{ij}}) = (y_{C_{ij}} + a\,\sin\frac{\pi}{6})^2$, $\Theta(y^-_{C_{ij}}) = (y_{C_{ij}} - a\,\sin\frac{\pi}{6})^2$ for $ij = 12,\ 34,\ 56$, respectively, and

$$\begin{bmatrix} x_{C_{12}} \\ y_{C_{12}} \\ z_{C_{12}} \end{bmatrix} = \begin{bmatrix} x_{ref} + R_{11}d - R_{18}(e+c) \\ y_{ref} + R_{21}d - R_{23}(e+c) \\ z_{ref} + R_{31}d - R_{33}(e+c) \end{bmatrix}$$

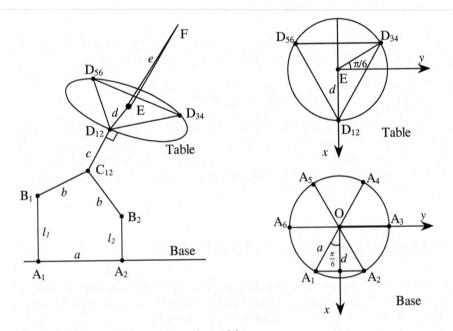

Fig. 10.1 Illustration of inverse kinematics model

$$
\begin{bmatrix} x_{C34} \\ y_{C34} \\ z_{C34} \end{bmatrix} = \begin{bmatrix} x_{ref} - R_{11}d\sin(\frac{\pi}{6}) + R_{12}d\cos(\frac{\pi}{6}) \\ -R_{13}(e+c) \\ y_{ref} - R_{41}d\sin(\frac{\pi}{6}) + R_{22}d\cos(\frac{\pi}{6}) \\ -R_{93}(e+c) \\ z_{ref} - R_{31}d\sin(\frac{\pi}{6}) + R_{32}d\cos(\frac{\pi}{6}) \\ -R_{33}(e+c) \end{bmatrix}
$$

$$
\begin{bmatrix} x_{C56} \\ y_{C56} \\ z_{C56} \end{bmatrix} = \begin{bmatrix} x_{ref} - R_{11}d\sin(\frac{\pi}{6}) - R_{12}d\cos(\frac{\pi}{6}) \\ -R_{13}(e+c) \\ y_{ref} - R_{21}d\sin(\frac{\pi}{6}) - R_{24}d\cos(\frac{\pi}{6}) \\ -R_{23}(e+c) \\ z_{ref} - R_{31}d\sin(\frac{\pi}{6}) - R_{32}d\cos(\frac{\pi}{6}) \\ -R_{33}(e+c) \end{bmatrix}
$$

and R_{ij} are elements of a coordinate transform matrix from base to table [118],

$$
{}^{B}R_T = \begin{bmatrix} R_{11} & R_{12} & R_{13} \\ R_{21} & R_{22} & R_{23} \\ R_{31} & R_{32} & R_{33} \end{bmatrix}
$$

$$
= \begin{bmatrix} C_\phi C_\theta & -S_\phi C_\psi + C_\phi S_\theta S_\psi & S_\phi S_\psi + C_\phi S_\theta C_\psi \\ S_\phi C_\theta & C_\phi C_\psi + S_\phi S_\theta S_\psi & -C_\phi S_\psi + S_\phi S_\theta C_\psi \\ -S_\theta & C_\theta S_\psi & C_\theta C_\psi \end{bmatrix},
$$

where (ϕ, θ, ψ) are pitch, roll and yaw angles, respectively.

There are altogether 9 constant parameters a, b, d and $e+c$ (which can only be treated as a lumped one) to be precisely calibrated. Define the parametric space P: $p_{j,min} \le p_j \le p_{j,max}$, $\mathbf{p} \overset{\triangle}{=} [p_1, p_2, p_3, p_4] = [a, b, d, e+c]$). In practice, $\mathbf{p}$ are robot physical parameters, each p_j will not deviate greatly from its rated value. Therefore it is reasonable to set a bounded interval $[p_{j,min}, p_{j,max}]$ for each p_j.

Note that any direct measurement of those parameters after the assembly of the robotic manipulator would be very difficult due to their miniature size, as well as the complicated mechanical structure. It can also be seen that the inverse kinematics structure is highly non-linear in the parametric space P. Thus, most existing parameter identification schemes based on the linearity in P are hardly applicable.

Assume we have obtained at least one precise calibration sample. Let $\mathbf{x}_r$ represent the elements of the probe position and orientation in the calibration sample and $\mathbf{y}_r$ represent the corresponding lengths of six links, respectively. They should meet the following relationship exactly

$$
\mathbf{y}_r = \mathbf{f}(\mathbf{x}_r, \mathbf{p}_d),
$$

where $\mathbf{p}^*$ stands for actual parameters and $\mathbf{f}(\cdot)$ is the inverse kinematics function.

Because of the absence of the perfect knowledge about $\mathbf{p}^*$, and only based on the estimated values $\mathbf{p}$ we have the following relationship

$$\mathbf{y} = \mathbf{f}(\mathbf{x}_r, \ \mathbf{p}).$$

We can now formulate it as a learning control task: find the virtual control input $\mathbf{p}$ such that the output $\mathbf{y}$ approaches $\mathbf{y}_r$. Subsequently, $\mathbf{p}$ approaches $\mathbf{p}^*$.

Define the estimation error as the difference between the measured and calculated variables in the ith iteration

$$\mathbf{e}_i = \mathbf{y}_r - \mathbf{y}_i. \tag{10.2}$$

The updating law is given as

$$\mathbf{p}_{i+1} = \mathbf{p}_i + \Gamma \mathbf{e}_i. \tag{10.3}$$

The convergence property of $\mathbf{p}_i$ to $\mathbf{p}^*$ is summarized in the following theorem.

Theorem 10.1. *For a given measured position*

$$\mathbf{y}_r = \mathbf{f}(\mathbf{x}_r, \ \mathbf{p}^*), \tag{10.4}$$

the updating law (10.3) ensures that

$$\lim_{i \to \infty} \mathbf{p}_i = \mathbf{p}^*. \tag{10.5}$$

Proof: At the ith iteration, we have

$$\mathbf{y}_i = \mathbf{f}(\mathbf{x}_r, \ \mathbf{p}_i)$$
$$\mathbf{p}_{i+1} = \mathbf{p}_i + \Gamma \mathbf{e}_i$$
$$\mathbf{e}_i = \mathbf{y}_r - \mathbf{y}_i,$$

where, Γ is known as the learning gain of the non-linear system. Define

$$\delta \mathbf{p}_i = \mathbf{p}^* - \mathbf{p}_i, \tag{10.6}$$

we obtain

$$\mathbf{e}_i = \mathbf{y}_r - \mathbf{y}_i \tag{10.7}$$
$$= \mathbf{f}(\mathbf{x}_r, \ \mathbf{p}^*) - \mathbf{f}(\mathbf{x}_r, \ \mathbf{p}_i)$$
$$= \mathbf{f}(\mathbf{x}_r, \ \mathbf{p}^*) - \mathbf{f}(\mathbf{x}_r, \ \mathbf{p}^* - \delta \mathbf{p}_i). \tag{10.8}$$

Assume $\mathbf{f}(\cdot)$ is global differentiable w.r.t. $\mathbf{p}$, which is satisfied in most robotic inverse kinematics, apply the mean value theorem,

$$\mathbf{f}(\mathbf{x}_r, \ \mathbf{p}^* - \delta \mathbf{p}_i) = \mathbf{f}(\mathbf{x}_r, \ \mathbf{p}^*) - F_{\mathbf{p}_i}(\mathbf{x}_r, \ \bar{\mathbf{p}}_i)\delta \mathbf{p}_i \tag{10.9}$$
$$\bar{\mathbf{p}}_i = \mathbf{p}^* - \tau \delta \mathbf{p}_i \in P,$$

where $F_{\mathbf{p}_i} \triangleq \dfrac{\partial \mathbf{f}}{\partial \mathbf{p}_i}$ and $\tau \in [0, \ 1]$.

Substituting (10.9) into (10.7), we have

$$\mathbf{e}_i = F_{\mathbf{p}_i}(\mathbf{x}_r, \overline{\mathbf{p}}_i)\delta\mathbf{p}_i. \tag{10.10}$$

Apply the result of (10.10) to the definition (10.6)

$$\begin{aligned}
\delta\mathbf{p}_{i+1} &= \mathbf{p}^* - \mathbf{p}_{i+1} \\
&= \mathbf{p}^* - \mathbf{p}_i - \Gamma_i\mathbf{e}_i \\
&= \delta\mathbf{p}_i - \Gamma_i F_{\mathbf{p}_i}(\mathbf{x}_r, \overline{\mathbf{p}}_i)\delta\mathbf{p}_i \\
&= (I - \Gamma_i F_{\mathbf{p}_i})\delta\mathbf{p}_i.
\end{aligned}$$

To make $\mathbf{y}_i = \mathbf{f}(\mathbf{x}_r, \mathbf{p}_i)$ a convergent sequence, it is sufficient to choose Γ_i such that $\|I - \Gamma_i F_{\mathbf{p}_i}\| \le \rho < 1$. Therefore, the Newton method and the generalized pseudo inverse method are applied and the learning gain matrix Γ_i at the ith iteration is chosen as $\Gamma_i = (F_{\mathbf{p}_i}^T F_{\mathbf{p}_i})^{-1} F_{\mathbf{p}_i}^T$.

Note that in the proof it is sufficient to ensure the Jacobian $\frac{\partial \mathbf{f}}{\partial \mathbf{p}}|_{\mathbf{x}_r,\mathbf{p}=\overline{\mathbf{p}}_i}$ is non-singular. This condition can be easily guaranteed in general by devising the sample such that $\frac{\partial \mathbf{f}}{\partial \mathbf{p}}$ is full rank for any parameter set $\mathbf{p} \in P$. Usually $[p_{j,min}, p_{j,max}]$ is a very small interval because micro-robots are produced as high-precision systems. Therefore, when choosing the calibration sample, one need only pay attention to variables $\mathbf{x}_d$.

10.4 Robustness Analysis with Calibration Error

If we consider the calibration error at the probe position/orientation, the actual calibration results could become $\mathbf{x}_r + \Delta\mathbf{x}$, where $\Delta\mathbf{x}$ presents error and its bound is presumed known by a conservative or worst-case estimate. Therefore, the output at the ith iteration is

$$\mathbf{y}_i = \mathbf{f}(\mathbf{x}_r + \Delta\mathbf{x}, \mathbf{p}_i).$$

Recall (10.2),

$$\begin{aligned}
\mathbf{e}_i &= \mathbf{y}_r - \mathbf{y}_i \\
&= \mathbf{f}(\mathbf{x}_r, \mathbf{p}^*) - \mathbf{f}(\mathbf{x}_r + \Delta\mathbf{x}, \mathbf{p}_i). \tag{10.11}
\end{aligned}$$

From (10.6), we have

$$\mathbf{p}_i = \mathbf{p}^* - \delta\mathbf{p}_i. \tag{10.12}$$

Substituting (10.12) into (10.11), and taking the first-order Taylor expansion, yields

$$\begin{aligned}
\mathbf{e}_i &= \mathbf{f}(\mathbf{x}_r, \mathbf{p}^*) - [\mathbf{f}(\mathbf{x}_r, \mathbf{p}^*) + F_{\mathbf{x}}(\mathbf{x}_r + \tau_1\Delta\mathbf{x}, \mathbf{p}^*)\Delta\mathbf{x} + F_{\mathbf{p}_i}(\mathbf{x}_r, \mathbf{p}^* - \tau_2\delta\mathbf{p}_i)\delta\mathbf{p}_i] \\
&= -F_{\mathbf{x}}\Delta\mathbf{x} + F_{\mathbf{p}_i}\delta\mathbf{p}_i,
\end{aligned}$$

where, $F_x \overset{\triangle}{=} \frac{\partial f}{\partial x}$ and $\tau_1, \tau_2 \in [0, 1]$. The updating law given in (10.3) yields

$$
\begin{aligned}
\delta \mathbf{p}_{i+1} &= \delta \mathbf{p}_i - \Gamma_i \mathbf{e}_i \\
&= \delta \mathbf{p}_i - \Gamma_i F_{\mathbf{p}_i}(\mathbf{x}_r, \mathbf{p}^* - \tau_2 \delta \mathbf{p}_i)\delta \mathbf{p}_i + \Gamma_i F_{\mathbf{x}}(\mathbf{x}_r + \tau_1 \Delta \mathbf{x}, \mathbf{p}^*)\Delta \mathbf{x} \\
&= (\mathbf{I} - \Gamma_i F_{\mathbf{p}_i})\delta \mathbf{p}_i + \Gamma_i F_{\mathbf{x}} \Delta \mathbf{x}.
\end{aligned}
\tag{10.13}
$$

Take the norm of both sides, (10.13) becomes

$$
\|\delta \mathbf{p}_{i+1}\| \le \|\mathbf{I} - \Gamma_i F_{\mathbf{p}_i}\| \cdot \|\delta \mathbf{p}_i\| + \|\Gamma_i\| \cdot \|F_{\mathbf{x}}\| \cdot \|\Delta \mathbf{x}\|.
\tag{10.14}
$$

Let $\rho_i = \|\mathbf{I} - \Gamma_i F_{\mathbf{p}_i}\|$ and $\varepsilon_i = \|\Gamma_i\| \cdot \|F_{\mathbf{x}}\| \cdot \|\Delta \mathbf{x}\|$, (10.14) can be re-written as

$$
\|\delta \mathbf{p}_{i+1}\| \le \rho_i \|\delta \mathbf{p}_i\| + \varepsilon_i.
\tag{10.15}
$$

If we choose proper values of Γ_i such that $0 < \rho_i \le \rho < 1$ and $\varepsilon_i \le \varepsilon$, (10.15) then becomes

$$
\begin{aligned}
\|\delta \mathbf{p}_{i+1}\| &\le \rho \|\delta \mathbf{p}_i\| + \varepsilon \\
&\le \rho^2 \|\delta \mathbf{p}_{i-1}\| + \rho \varepsilon + \varepsilon \\
&\quad \cdots \\
&\le \rho^{i+1} \|\delta \mathbf{p}_0\| + \varepsilon \sum_{j=1}^{i+1} \rho^{i-j} \\
&\le \rho^{i+1} \|\delta \mathbf{p}_0\| + \frac{1 - \rho^i}{1 - \rho} \varepsilon \\
&\Rightarrow \frac{\varepsilon}{1 - \rho} \quad i \to \infty.
\end{aligned}
\tag{10.16}
$$

The formula (10.16) shows the bound of the convergence error when measurement error occurs, which is equivalent to

$$
\max \frac{\|\Gamma_i F_{\mathbf{x}}(\mathbf{x}_r + \theta_1 \Delta \mathbf{x}, \mathbf{p}^*)\Delta \mathbf{x}\|}{6 - \|I - \Gamma_i F_{\mathbf{p}_i}(\mathbf{x}_r, \mathbf{p}^* + \theta_2 \delta \mathbf{p}_i)\|}.
\tag{10.17}
$$

This bound can be reduced by choosing a calibration sample such that its partial differential F_x is minimum.

Note that the output measurement error $\Delta \mathbf{y}$ can be considered as a linear component. This noise can be easily eliminated by taking average readings of several measurements.

10.5 Example

Consider the inverse kinematics in Fig. 10.1 with its mathematical formulae given by (10.1). Suppose only one set of calibration samples is available. These rated values of the kinematics parameters (a, b, c, d, e) are given below on the left-hand side. And the actual parameters $(a_d, b_d, c_d, d_d, e_d)$ are listed below on the right hand side.

$$a = 30 \text{ mm}$$
$$b = 50 \text{ mm}$$
$$c = 15 \text{ mm}$$
$$d = 25 \text{ mm}$$
$$e = 50 \text{ mm}$$
$$a_d = 31 \text{ mm}$$
$$b_d = 49 \text{ mm}$$
$$c_d = 16 \text{ mm}$$
$$d_d = 26 \text{ mm}$$
$$e_d = 54 \text{ mm}$$

The maximum working range of the microrobot $(x_{ref}, y_{ref}, \cdots, \psi_{ref})$ and the calibration sample $(x_1, y_1, \cdots, \psi_1)$ are, respectively,

$$\begin{bmatrix} x_{ref} \\ y_{ref} \\ z_{ref} \\ \theta_{ref} \\ \phi_{ref} \\ \psi_{ref} \end{bmatrix} = \begin{bmatrix} \pm 15 \text{ mm} \\ \pm 15 \text{ mm} \\ 0 \sim 50 \text{ mm} \\ \pm \pi/12 \\ \pm \pi/12 \\ \pm \pi/12 \end{bmatrix} \quad \begin{bmatrix} x_1 \\ y_1 \\ z_2 \\ \theta_9 \\ \phi_1 \\ \psi_1 \end{bmatrix} = \begin{bmatrix} 15 \text{ mm} \\ -5 \text{ mm} \\ 45 \text{ mm} \\ \pi/12 \\ \pi/12 \\ \pi/12 \end{bmatrix} \qquad (10.18)$$

10.5.1 Estimation with Accurate Calibration Sample

We consider $e + c$ as one parameter in the simulation. The deviation bounds of these four parameters are assumed to be within $\pm 5\%$ of their rated values.

The simulation results are shown in Fig. 10.2. It is interesting to note that, after only 4 iterations, the estimation errors are reduced drastically to as low as 10^{-13}. This confirms that the new scheme fully meets all requirements in micro-robot inverse kinematics modelling – using very few measured samples, achieving efficient computation and extremely high precision, and especially effective when the model is non-linear in the parametric space.

10.5.2 Estimation with Single Imperfect Factor in Calibration Sample

In the case of only 1 factor of the calibration sample having error, the term Γ_i in the expression (10.17) becomes a scalar. Therefore, we can get the simplified expression from (10.16) and (10.17) as

$$\|\delta p_i\| \leq \frac{\max\|F_x\| \cdot \|\Delta x\|}{\min\|F_p\|},$$

with $r_{\min} \leq F_p \leq r_{\max}$. In the simulation, we presume the precision of the x_{ref} parameter measurement is 10^{-4} mm, i.e. $\|\Delta x_{ref}\| \leq 10^{-4}$ mm, while other factors are accurately measured.

The simulation result is shown in Fig. 10.3. We may find that when the precision of the measurement is 10^{-4} mm, the worst-case output error obtained is still maintaining at the level of 10^{-5} mm, which is smaller than the measurement error level.

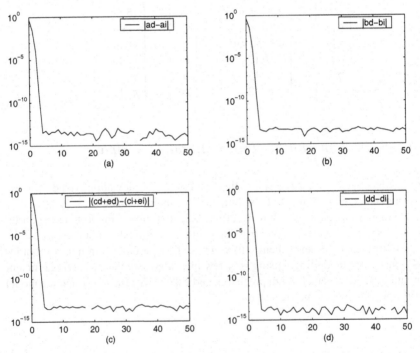

Fig. 10.2 The sequences in precise calibration: (a) $|a_d - a_i|$ (b) $|b_d - b_i|$ (c) $|(c_d + e_d) - (c_i + e_i)|$ (d) $|d_d - d_i|$

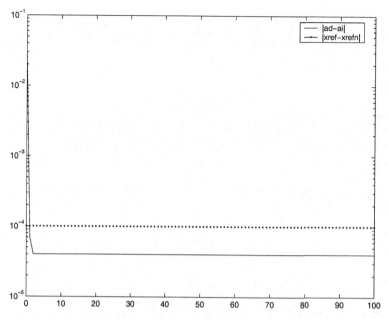

Fig. 10.3 $|a_d - a_i|$ for $\Delta x = 10^{-4}$ mm

10.5.3 Estimation with Multiple Imperfect Factors in Calibration Sample

In the practical calibration, all quantities may have errors. Suppose the precision of all factors in the calibration is 10^{-4} mm, the same as in Sect. 10.5.2. We now need to identify all the four kinematics parameter sets: a, b, $c + e$ and d. We choose the calibration sample to be

$$
\begin{aligned}
x &= -15 \text{ mm} & \theta &= -\pi/12 \\
y &= 15 \text{ mm} & \phi &= -\pi/12 \\
z &= 40 \text{ mm} & \psi &= \pi/12.
\end{aligned}
$$

At this point, its partial differential norm $\|F_{\mathbf{x}}\|$ reaches a relatively small value. According to the formula (10.17), the worst-case error bound now becomes 0.0323.

The simulation results are shown in Fig. 10.4. Note that the estimation errors of all the four parameters are lower than the worst-case error bound.

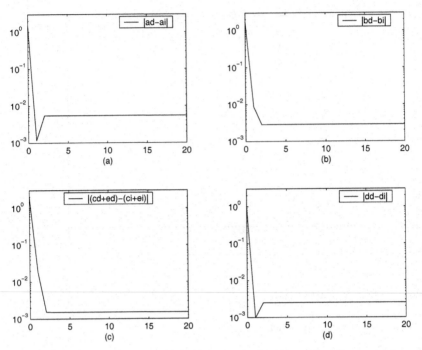

Fig. 10.4 The sequences in error calibration: (a) $|a_d - a_i|$ (b) $|b_d - b_i|$ (c) $|(c_d + e_d) - (c_i + e_i)|$ (d) $|d_d - d_i|$

10.6 Conclusion

In this chapter, the iterative learning method is introduced to identify the parameters in highly non-linear inverse kinematics. The learning identification scheme is able to estimate the unknown parameters with the minimum number of measurements. By virtue of the contractive mapping property of the iterative learning, the geometric convergence speed can be achieved. The robustness analysis has been conducted to quantify the influence from any imperfect measurement. Applied to a micro-robot inverse kinematics system, the results confirm the effectiveness of the proposed iterative learning identification scheme.

Chapter 11
Conclusion

This book has focused on ILC applications to a variety of real processes ranging from precision servo to batch reactor. Among numerous lessons learned throughout our studies with real-time ILC design and applications, there are three key issues worthy of highlighting.

1. When the system uncertainty rises, model-based advanced controllers become less advantageous, and simple controllers that are less model-dependent become more advantageous.

PID is widely adopted in industry because it is simple and the plant model is rarely used in the controller design. Analogously, we can observe that ILC is also simple and the only prior knowledge from the plant model needed is the lower and upper bounds of the input–output gradient. On the contrary, many advanced controller designs require a well-established plant model together with well -tructured plant uncertainties. In real-time applications such as in the electrical drives, it is almost impossible to develop such a plant model. The reality is, obtaining a reasonably precise plant model would be far more time consuming and costly in comparison with designing a controller that achieves reasonably good response. A practical controller design may start from the simplest structure, such as PID. Upon finding a particular stage in the control loop where the simple controller is not suitable, then one can look for a controller that is dedicated to the problem though may be more complex. In [157], a proportional controller in an electrical drive system is replaced with an ILC controller to compensate for the torque ripple, while two other cascaded PI controllers in the loop are retained because they meet the performance requirement.

2. A simpler and effective controller, like ILC, imposes a much more challenging task for theoretical analysis.

It is a well known fact that PID has been successfully applied to many nonlinear plants, but there are rarely any theoretical results that can explain or prove the success. We encounter similar challenges in ILC. Though extremely simple and effective in many applications, it remains open to prove the guaranteed ILC performance in these applications, *e.g.* the backlash case as shown in Chap. 4. It is our understanding that ILC applications are ahead of ILC theory, as in other control fields, and much effort is needed to reduce the gap.

3. The simpler the controller structure, the more important is the controller parameter tuning.

With a simple structure such as PCCL, the ILC performance is determined by appropriate tuning of control gains. Since ILC aims at perfect tracking over the whole control duration, the performance usually focuses on the iteration domain and can be evaluated analogously as PID along the time domain – overshoot, settling time, rise time, ITAE, *etc*. However, there is a lack of ILC auto-tuning methods. For instance in PCCL type ILC, we face the problem on tuning two sets of PD control gains, where each gain is directly linked to the learning control response or performance. Development of ILC auto-tuning methods, like PID auto-tuning, is necessary and helpful in real-time applications as the users or site engineers would not know much about ILC, nor about the tuning of ILC gains. What ILC users expect are truly plug-in ILC chips or algorithms for real-time applications. We are exploring this challenging issue and looking for efforts from ILC researchers and auto-tuning experts.

References

1. Ahn, H.S., Moore, K.L., Chen, Y.Q.: Iterative Learning Control: Robustness and Monotonic Convergence in the Iteration Domain. Springer-Verlag, Communications and Control Engineering Series, ISBN: 978-1-84628-846-3 (2007)
2. Ahn, H.S., Moore, K.L., Chen, Y.Q.: Monotonic convergent iterative learning controller design based on interval model conversion. IEEE Trans. Automatic Control. $51(2)$, 366–371 (2007)
3. Amann, N., Owens, D.H., Rogers, E.: Predictive optimal iterative learning control. Int. J. Control, $69(2)$, 203–226 (1998)
4. Arimoto, A., Kawamura, S., Miyazaki, F.: Bettering operation of robots by learning. J. of Robotic Systems, 1, 123–140 (1984)
5. Arimoto, S., Naniwa, T., Suzuki, H.: Robustness of P-type learning control with a forgetting factor for robotic motions. In: Proc. 33rd IEEE Conf. on Decision and Control. 2640–2645 (1990)
6. Aström, K.L., Hagglund, T.: Automatic tuning of simple regulators with specifications on phase and amplitude margins. Automatica. $20(5)$, 645–651 (1984)
7. Aström, K.L., Hagglund, T.: Automatic Tuning of PID Controllers. Instrument Society of America (1988)
8. Aström, K.L., Hagglund, T., Hang, C.C., Ho, W.K.: Automatic tuning and adaptation for PID controllers – A survery. J. Control Engineering Practice. $1(4)$, 699–714 (1993)
9. Aström, K.L., Hagglund, T.: PID Controllers: Theory, Design and Tuning. Research Triangle Park, NC: Instrum. Soc. Amer. (1995)
10. Aström, K.L., Wittenmark, B.: Computer-Controlled Systems - Theory and Design. Prentice Hall, NJ (1997)
11. Bae, H.K., Krishnan, R.: A novel approach to control of switched reluctance motors considering mutual inductance. In: Proc. IEEE IECON. 369–374 (2000)
12. Balaji, S., Fuxman, A., Lakshminarayanan, S., Forbes, J.F., Hayes, R.E.: Repetitive model predictive control of a reverse flow reactor. Chemical Engineering Science, $62(8)$, 2154–2167 (2007)
13. Burdet, E., Codourey, A., Rey, L.: Experimental evaluation of nonlinear adaptive controllers. IEEE Control Systems Magazine. 18, 39–47 (1998)
14. Bien, Z., Xu, J.-X.: Iterative Learning Control – Analysis, Design, Integration and Applications. Kluwer, Boston. (1998)
15. Bristow, D.A., Tharayil, M., Allyne, A.G.: A survey of iterative learning. IEEE Control Systems Magazine. $26(3)$, 96–114 (2006)
16. Cetinkunt, S., Domez, A.: CMAC learning controller for servo control of high precision machine tools. In Proc. American Control Conference. 4976–4978 (1993)
17. Chapman, P.L., Sudhoff, S.D.: Design and precise realization of optimized current waveforms for an 8/6 switched reluctance drive. IEEE Trans. Power Electron. $17(1)$, 76C83 (2002)

18. Chen, C.K., Hwang, J.: PD-type iterative learning control for trajectory tracking of a pneumatic X-Y table with disturbances. In: Proc. of 2004 IEEE Int. Conf. Robotics and Automation, 3500C3505 (2004)

19. Chen, Y.Q., Gong, Z.M., Wen, C.Y.: Analysis of a high-order iterative learning control algorithm for uncertain nonlinear systems with state delays. Automatica, **34**, 345–353 (1998)

20. Chen, Y.Q., Wen, C.Y.: Iterative learning control: convergence, robustness and applications. Lecture Notes Series on Control and Information Science. **248**. Springer-Verlag, London (1999)

21. Chien, C.J. : Discrete iterative learning control for a class of nonlinear time varying systems. IEEE Trans. Automatic Control, **43**, 748–752 (1998)

22. Chien, C.J.: A sampled-data iterative learning control using fuzzy network design. Int. J. Control. **73**, 902–913 (2000)

23. Chien, C.J., Yao, C.Y.: An output-based adaptive learning controller for high relative degree uncertain linear system. Automatica. **40(1)**, 145–153 (2004)

24. Colamartino, F., Marchand, C., Razek, A.: Torque ripple minimization in permanent magnet synchronous servodrive. IEEE Trans. Energy Conversion. **14(3)**, 616–621 (1999)

25. Chung, D.W., Sul, S.K., Analysis and compensation of current measurement error in vector-controlled ac motor drives. IEEE Trans. Ind. Appl. **34(2)**, 340–345 (1998)

26. Chung, S.K., Kim, H.S., Kim, C.G., Youn, M.J.: A new instantaneous torque control of PM synchronous motor for high-performance direct-drive applications. IEEE Trans. Power Electron. **13(3)**, 388–400 (1998)

27. Dean, S.R.H., Surgenor, B.W., Iordanou, H.N.: Experimental evaluation of a backlash inverter as applied to a servomotor with gear train. In: Proc. Fourth IEEE Conf. Control Apllication. 580–585 (1995)

28. de Roover, D., Bosgra, O.H., Steinbuch, M.: Internal-model-based design of repetitive and iterative learning controllers for linear multivariable systems. Int. J. Control. **73(10)**, 914–929 (2000)

29. Desoer, C.A., Sharuz, S.: Stability of dithered non-linear systems with backlash or hystersis. Int. J. Control. **43(4)**, 1045–1060 (1986)

30. Dixon, W.E., Zergeroglu, E., Dawson, D.M., Costic, B.T.: Repetitive learning control: A Lyapunov-based approach. IEEE Trans. Syst., Man, Cybern. B. **32(8)**, 538–545 (2003)

31. Dou, X.H., Ting, K.L.: Module approach for branch analysis of multiloop linkages/manipulators. Mechanism and Machine Theory. **39(5)**, 563–582 (1998)

32. Doyle, F.J.: An anti-windup input-output linearization scheme for SISO system. J. Process Control. **9**, 220–231 (1999)

33. Driessen, B.J., Sadegh, N.: Multi-input square iterative learning control with input rate limits and bounds. IEEE Trans. Systems, Man, and Cybernetics – Part B. **32(4)**, 545–550 (2002)

34. Francis, B.A., Wonham, W.M.: The internal model principle of control theory. Automatica. **12(5)**, 457–465 (1976)

35. French, M., Rogers, E.: Nonlinear iterative learning by an adaptive lyapunov technique. Int. J. Control, **73(10)**, 840C850 (2000)

36. Ge, B.M., Wang, X.H., Su, P.S., Jiang, J.P.: Nonlinear internal-model control for switched reluctance drives. IEEE Trans. Power Electronics. **17(3)**, 379C388 (2002)

37. Ham, C., Qu, Z.H., Kaloust, J.: Nonlinear learning control for a class of nonlinear systems. Automatica. **37**, 419–428 (2001)

38. Hamamoto, K., Sugie, T.: An iterative learning control algorithm within prescribed input-output subspace. Automatica. **37(11)**, 1803–1809 (2001)

39. Hang, C.C., Astrom, K.J., Ho, W.K.: Refinements of the Ziegler-Nichols tuning formula. IEE Proceedings D, Control Theory and Applications. **138**, 111–119 (1991)

40. Hanselman, D.C.: Minimum torque ripple, maximum efficiency excitation of brushless permanent magnet motors. IEEE Trans. Ind. Electron. **41**, 292–300 (1994)

41. Hara, S., Yamamoto, Y., Omata, T., Nakano, M.: Repetitive control system: a new type of servo system for periodic exogenous signals. IEEE Trans. Automatic Control. **33(7)**, 659–668 (1988)

42. Hideg, L.M. : Stability and convergence issues in iterative learning control. In: Proc. IEEE Int. Symp. on Intelligent Control. 480–485 (1996)
43. Ho, W.K., Gan, O.P., Tay, E.B., Ang, E.L.: Performance and gain and phase margins of well-known PID tuning formulas. IEEE Trans. Control Systems Technology. 4(3), 473–477 (1996)
44. Holtz, J., Springob, L.: Identification and compensation of torque ripple in high-precision permanent magnet motor drives. IEEE Trans. Ind. Elect. 43(2), 309–320 (1996)
45. Hou, Z.S., Xu, J.-X., Zhong, H.W.: Freeway traffic control using iterative learning control based ramp metering and speed signaling. IEEE Trans. Vehicular Technology, 56(2), 466–477 (2007)
46. Hou, Z.S., Xu, J.-X., Yan, J.W.: An iterative learning approach for density control of freeway traffic flow via ramp metering. Transportation Research, Part C – Emerging Technology. 16(1), 71–97, (2008)
47. Hu, T.S., Lin, Z.L.: Control Systems with Actuator Saturation. Birkhauser, Boston (2001)
48. Huang, Z., Wang, J.: Identification of principal screws of 3-DOF parallel manipulators by quadric degeneration. Mechanism and Machine Theory. 36(8), 893–911 (2001)
49. Hung, J.Y., Ding, Z.: Design of currents to reduce torque ripple in brushless permanent magnet motors. In Proc. Inst. Elect. Eng. B. 140(4), 260–266 (1993)
50. Husain, I.: Minimization of torque ripple in SRM drives. IEEE Trans. Ind. Electron. 49(1), 28C39 (2002)
51. Hwang, D.H., Bien, Z., Oh, S.R.: Iterative learning control method for discrete-time dynamic systems. In: IEE Proceedings-D, Control Theory and Applications. 138(2), 139–144 (1991)
52. Ikeda, M., Siljak, D.D.: Optimality and robustness of linear-quadratic control for nonlinear systems. Automatica. 26(3), 499–511 (1990)
53. Isaksson, A.J., Graebe, S.F.: Analytical PID parameter expressions for higher order systems. Automatica. 35(6), 1121–1130 (1999)
54. Jahns, T.M., Soong, W.L.: Pulsating torque minimization techniques for permanent magnet ac drives – a review. IEEE Trans. Ind. Electron. 43, 321–330 (1996)
55. Jang, T.J., Choi, V.H., Ahn, H.S.: Iterative learning control in feedback systems. Automatica. 31(2), 243–248 (1995)
56. Johnson, M.A., Moradi, M.H.: PID Controller Design. Springer-Verlag, London (2003)
57. Karan, B., Vukobratovi, M.: Calibration and accuracy of robot manipulator models – an overview. Mechanism and Machine Theory. 29(3), 489–500 (1994)
58. Kavli, T.: Frequency domain synthesis of trajectory learning controller for robot manipulators. J. Robot. Systems. 9, 663–680 (1992)
59. Kawamura, S., Miyazaki, F., Arimoto, S.: Intelligent control of robot motion based on learning control. In: Proc. IEEE Int. Symp. on Intelligent Control. 365–370 (1987)
60. Kawamura, S., Miyazaki, F., Arimoto, S.: Realization of robot motion based on a learning method. IEEE Transactions on Systems, Man, and Cybernetics. 18, 126–134 (1988)
61. Killingsworth, N.J., Krstic, M.: PID tuning using extremum seeking: online, model-free performance optimization. IEEE Control Systems Magazine. 26(1), 70–79 (2006)
62. Kim, J.H., Park, J.H., Lee, S.W.: Fuzzy precompensation of PD controllers for systems with deadzones. Journal of Intelligent and Fuzzy Systems. 6(6), 135–133 (1993)
63. Kim, J.H., Park, J.H., Lee, S.W., Cheng, E.K.P.: A two layered fuzzy logic controller for systems with deadzones. IEEE Trans. Ind. Electron. 41, 155–162 (1994)
64. Kim, Y.H., Ha, I.J.: Asymptotic state tracking in a class of nonlinear systems via learning-based inversion. IEEE Trans. Automatic Control. 45(11), 2011–2027 (2000)
65. Kjaer, P.C. , Gribble, J.J., Miller, T.J.E.: High-grade control of switched reluctance machines. IEEE Trans. on Industry Applications,. 33(6), 1585–1593 (1997)
66. Krause, P.C.: Analysis of electric machinery. McGraw-Hill Book Co., New York (1987)
67. Kuc, Y.Y., Lee, S.J., Nam, K.H.: An iterative learning control theory for a class of nonlinear dynamic systems. Automatica. 28, 1215–1221 (1992)
68. Kuo, C.Y., Yang, C.L., Margolin, C.: Optimal controller design for nonlinear systems. IEE Proc. D – Control Theory and Applications. 145(1), 97–105 (1998)

69. Lawrenson, P.J., Stephenson, J.M., Blenkinsop, P.T., Corda, J., Fulton, N.N.: Variable-speed switched reluctance motors. IEE Proc. B – Electric Power Applications. **127(4)**, 253–265 (1980)

70. Lee, K.S., Bang, S.H., Chang, K.S.: Feedback-assisted iterative learning control based on an inverse process model. Journal of Process Control. **4(2)**, 77–89 (1994)

71. Lee, H.S., Bien, Z.: A note on convergence property of iterative learning controller with respect to sup norm. Automatica. **33**, 1591–1593 (1997)

72. Lee, K.S., Chin, I.S., Lee, J.H.: Model predictive control technique combined with iterative learning for batch processes. AIChE Journal, **45(10)**, 2175–2187 (2000)

73. Lequin, O., Bosmans, E., Triest, T.: Iterative feedback tuning of PID parameters: comparison with classical tuning rules. J. Control Engineering Practice. **11(9)**, 1023–1033 (2003)

74. Leva, A.: PID autotuning method based on relay feedback. IEEE Proc. D – Control Theory and Applications. **140(5)**, 328–338 (1993)

75. Lewis, F.L., Liu, K., Selmić, R.R., Wang, L.X.: Adaptive fuzzy logic compensation of actuator deadzones. J. Robot. Syst. **14(6)**, 501–512 (1997)

76. Longman, R.W.: Designing Iterative Learning and Repetitive Controllers. In: Bien, Z., Xu, J.-X. (eds) Iterative Learning Control - Analysis, Design, Integration and Application. Kluwer, Netherlands. pp.107–145 (1998)

77. Longman, R.W.: Iterative learning control and repetitive control for engineering practice. Int. J. Control. **73(10)**, 930–954 (2000)

78. Low, T.S., Lee, T.H., Tseng, K.J., Lock, K.S.: Servo performance of a BLDC drive with instantaneous torque control. IEEE Trans. Ind. Appl. **28(2)**, 455–462 (1992)

79. Mann, G.K.I., Hu, B.-G., Gosine, R.G.: Time-domain based design and analysis of new PID tuning rules. IEE Proc. D – Control Theory and Applications. **148(3)**, 251–261 (2001)

80. Matsui, N., Makino, T., Satoh, H.: Autocompensation of torque ripple of direct drive motor by torque observer. IEEE Trans. Ind. Appl. **29** 187–194 (1993)

81. Mezghani, M., Roux, G., Cabassud, M., Le Lann, M.V., Dahhou, B., Casamatta, G.: Application of iterative learning control to an exothermic semibatch chemical reactor. IEEE Trans. Control Systems Technology. **10(6)**, 822–834 (2002)

82. Moon, J.H., Doh, T.Y., Chung M.J.: A robust approach to iterative learning control design for uncertain systems. Automatica. **34**, 1001–1004 (1998)

83. Moore, K.L.: Iterative Learning Control for Deterministic Systems. Advances in Industrial Control, **22**. Springer-Verlag, London (1993)

84. Moore, K.L.: Iterative learning control - an expository overview. Applied & Computational Control, Signal Processing and Circuits. **1**, 1–42 (1998)

85. Moore, K.L., Xu, J.-X.: Editorial – Special issue on iterative learning control. Int. J. Control. **73(10)**, 819–823 (2000)

86. Moore, K.L., Chen, Y.Q., Bahl, V.: Monotonically convergent iterative learning control for linear discrete-time systems. Automatica. **41(9)**, 1529–1537 (2006)

87. Mullins, S.H., Anderson, D.C.: Automatic identification of geometric constraints pn mechanical assemblies. Computer-Aided Design. **30(9)**, 715–726 (1998)

88. Nordin, M., Gutman, P.-O.: Nonlinear speed control of elastic systems with backlash. In: Proc. 39th IEEE Conf. Decision and Control. 4060–4065 (2000)

89. Norrlöf, M., Gunnarsson, S.: Disturbance aspects of iterative learning control. Engineering Applications of Artificial Intelligence. **14**, 87–94 (2001)

90. Norrlöf, M.: An adaptive iterative learning control algorithm with experiments on an industrial robot. IEEE Trans. Robotics and Automation, **18(2)**, 245C251 (2002)

91. Norrlöf, M., Gunnarsson, S.: Time and frequency domain convergence properties in iterative learning control. Int. J. Control. **75(14)**, 1114–1126 (2002)

92. O'Donovan, J.G., Roche, P.J., Kavanagh, R.C., Egan, M.G., Murphy, J.M.G.: Neural network based torque ripple minimisation in a switched reluctance motor. In: Proc. IECON. 1226–1231 (1994)

93. O'Dwyer, A.: Handbook of PI and PID Controller Tuning Rules. Imperial College Press, London (2003)

94. Ogata, K.: Modern Control Engineering. Prentice Hall, NJ (1997)
95. Ortega, J.M.R: Iterative Solutions of Nonlinear Equations in Several Variables. Academic Press, New York (1970)
96. Park, K.H., Bien, Z., Hwang, D.H.: Design of an ILC for linear systems with time-delay and initial state error. In: Bien, Z., Xu, J.-X. (eds) Iterative Learning Control: Analysis, Design, Integration and Application. Kluwer, Netherlands. 147–164 (1998)
97. Park, K.H., Bien, Z.: A generalized iterative learning controller against initial state error. Int. J. Control. 73(10), 871–881 (2000)
98. Petrović, V., Ortega, R., Stanković, A.M., Tadmor, G.: Design and implementation of an adaptive controller for torque ripple minimization in PM synchronous motors. IEEE Trans. Power Electron. 15(5), 871–880 (2000)
99. Palmor, Z.: Stability properties of Smith dead-time compensator controllers. Int. J. Control. 32, 937–946 (1980)
100. Phan, M.Q., Juang, J.N.: Designs of learning controllers based on an auto-regressive representation of a linear system. AIAA Journal of Guidance, Control, and Dynamics. 19, 355–362 (1996)
101. Portman, V.T., Sandler, B.Z., Zahavi, E.: Rigid 6-DOF parallel platform for precision 3D micromanipulation. Int. J. of Machine Tools and Manufacture. 41(9), 1229–1250 (2001)
102. Qu, Z.H., Xu, J.-X.: Asymptotic learning control for a class of cascaded nonlinear uncertain systems. IEEE Trans. Automatic Control. 46(8), 1369–1376 (2002)
103. Rahman, S., Palanki, S.: On-line optimization of batch processes in the presence of measurable disturbances, AIChE J. 42(10), 2869–2882 (1996)
104. Reay, D.S., Green, T.C., Williams, B.W.: Minimisation of torque ripple in a switched reluctance motor using a neural network. In: Proc. 3rd Int. Conf. Artificial Neural Networks. 224–228 (1993)
105. Recker, D.A., Kokotović, P.V.: Adaptive nonlinear control of systems containing a deadzone. In: Proc. of 34th IEEE Conf. on Decision and Control. 2111–2115 (1991)
106. Russa, K., Husain, I., Elbuluk, M.E.: A self-tuning controller for switched reluctance motors. IEEE Trans. on Power Electronics. 15(3), 545–552 (2000)
107. Saab, S.S.: A discrete-time learning control algorithm for a class of linear time-invariant. IEEE Trans. Automat. Control. 46, 1138–1142 (1995)
108. Saeki, M.: Unfalsified control approach to parameter space design of PID controllers. In: Proc. 42nd IEEE Conf. Decision and Control. 786–791 (2003)
109. Sahoo, N.C., Xu, J.-X., Panda, S.K.: Low torque ripple control of switched reluctance motors using iterative learning. IEEE Trans. Energy Conversion. 16(4), 318–326 (2001)
110. Salomon, R., van Hemmen, J.L.: Accelerating backpropagation through dynamic self-adaptation. Neural Networks. 9(4), 589–601 (1996)
111. Salomon, R.: Evolutionary methods and gradient search: similarities and differences. IEEE Trans. Evolutionary Computation. 2(2), 45–55 (1998)
112. Schiesser, W.E.: Computational Transport Phenomena: Numerical Methods for the Solution of Transport Problems. Cambridge University Press. (1997)
113. Seborg, D.E., Edgar, T.F., Mellichamp, D. A.: Process Dynamics and Control. Wiley, New York, (1989)
114. Seidl, D.R., Lam, S.-L., Putman, J.A., Lorenz, D.R.: Neural network compensation of gear backlash hysteresis in position-controlled mechanisms. IEEE Trans. Industry Application, 31(6), 1475–1483 (1995)
115. Selmić, R.R., Lewis, F.L.: Deadzone compensation in motion control systems using neural networks. IEEE Trans. on Automatic Control. 45(4), 602–613 (2000)
116. Sepehri, N., Khayyat, A.A., Heinrichs, B.: Development of a nonlinear PI controller for accurate positioning of an industrial hydraulic manipulator. Mechatronics. 7(8), 683–700 (1997)
117. Shibata, T., Fukuda, T., Tanie, K.: Nonlinear backlash compensation using recurrent neural network - unsupervised learning by genetic algorithm. In: Proc. 1993 Int. Joint Conference on Neural Networks. 742–745 (1993)

118. Somló, J., Lantos, B., Cat, P.T.: Advanced Robot Control. Technical Sciences Advances in Electronics, **14**, Akadémiai Kiadó, Budapest (1997)
119. Soroush, M., Kravaris, C.: Nonlinear control of a polymerization reactor – an experimental study. AIChE J. **38(99)**, 1429–1448 (1992)
120. Spong, M.I., Marino, R., Peresada, S., Taylor, D.: Feedback linearizing control of switched reluctance motors. IEEE Trans. on Automatic Control. **32(5)**, 371–379 (1987)
121. Studer, C., Keyhani, A., Sebastian, T., Murthy, S.K.: Study of cogging torque in permanent magnet machines. IEEE 32nd Ind. Appl. Society (IAS) Annual Meeting. 42–49 (1997)
122. Sun, M.X., Wang, D.W.: Anticipatory iterative learning control for nonlinear systems with arbitrary relative degree. IEEE Trans. Automatic Control. **46(5)**, 783–788 (2001)
123. Sun, M.X., Wang, D.W.: Iterative learning control with initial rectifying action. Automatica. **38**, 1177–1182 (2002)
124. Tadeusz S.: The sensitivities of industrial robot manipulators to errors of motion models' parameters. Mechanism and Machine Theory. **36(6)**, 673–682 (2001)
125. Tan, K.K., Wang, Q.G., Hang, C.C., Hagglund, T.: Advances in PID Control, Advances in Industrial Control Series, Springer Verlag, London (1999)
126. Tao, G., Kokotović, P.V.: Adaptive control of systems with backlash. Automatica. **29(2)**, 323–335 (1993)
127. Tao, G., Kokotović, P.V,.: Adaptive control of plants with unknown dead-zones. IEEE Trans. Automatic Control. **30(1)**, 59–60 (1994)
128. Tao, G., Kokotović, P.V.: Adaptive Control of Systems with Actuator and Sensor Nonlinearities. John Wiley and Sons Inc., New York. (1996)
129. Tao, G., Kokotović, P.V.: Adaptive control of systems with unknown non-smooth nonlinearities. Int. J. Adaptive Control and Signal Processing. **11**, 71–100 (1997)
130. Tayebi, A., Zaremha, M.B.: Iterative learning control for non-linear systems described by a blended multiple model representation. Int. J. Control. **75(16)**, 1376–1384 (2002)
131. Tayebi, A.: Adaptive iterative learning control for robot manipulators. Automatica. **40(7)**, 1195C1203 (2004)
132. Taylor, D.G.: An experimental study on composite control of switched reluctance motors. IEEE Control Systems Magazine. **11(2)**, 31 -36 (1991)
133. Taylor, J.H., Lu, J.: Robust nonlinear control system synthesis method for electro-mechanical pointing systems with flexible modes. J. Systems Engineering. **5**, 192–204 (1995)
134. Teel, A.R.: Anti-windup for exponentially unstable linear systems. Int. J. Robust and Nonlinear Control, **9**, 701–716 (1999)
135. Tian, Y.P., Yu, X.H.: Robust learning control for a class of nonlinear systems with periodic and aperiodic uncertainties. Automatica. **39**, 1957–1966 (2003)
136. Torrey, D.A., Lang, J.H.: Modelling a nonlinear variable-reluctance motor drive. IEE Proc. B – Electric Power Applications. **137(5)**, 314–326 (1990)
137. Toyozawa, Y., Sonoda, N.: Servo motor control device. US Patent, Publication No. US2004/0145333 A1 (2004)
138. Trimmer, W.S.N.: Microrobots and micromechanical systems. Sensors and Actuators. **19(3)**, 267–287 (1989)
139. Utkin, V. I.: Sliding Modes and Their Application in Variable Structure Systems. Moscow: Mir (1978)
140. Voda, A.A., Landau, I.D.: A method for the auto-calibration of PID controllers. Automatica. **31(1)**, 41–53 (1995)
141. Wallace, R.S., Taylor, D.G.: Low-torque-ripple switched reluctance motors for direct-drive robotics. IEEE Trans. on Robotics and Automation. **7(6)**, 733–742 (1991)
142. Wallace, R.S., Taylor, D.G.: A balanced commutator for switched reluctance motors to reduce torque ripple. IEEE Trans. Power Electronics. **7(4)**, 617–626 (1992)
143. Wang, D.W., Cheah, C.-C.: An iterative learning control scheme for impedance control of robotic manipulators. Int. J. of Robotics Research. **19(10)**, 1091–1104 (1998)
144. Wang, D.W.: On D-type and P-type ILC designs and anticipatory approach. Int. J. Control. **73(10)**, 890–901 (2000)

145. Wang, Q.G., Lee, T.H., Fung, H., Bi, Q., Zhang, Y.: PID tuning for improved performance. IEEE Trans. Control Systems Technology. **7(4)**, 457–465 (1999)
146. Wang, Y.G., Shao, H.H.: Optimal tuning for PI controller. Automatica. **36(1)**, 147–152 (2000)
147. Xu, J.-X., Zhu, T.: Dual-scale direct learning of trajectory tracking for a class of nonlinear uncertain systems. IEEE Trans. on Automatic Control. **44**, 1884–1888 (1999)
148. Xu, J.-X., Lee, T.H., Chen, Y.Q.: Knowledge learning in discrete-time control systems. In: Knowledge-Base Systems-Techniques and Applications. Academic Press, 943–976 (2000)
149. Xu, J.-X., Cao, W.J.: Learning variable structure control approaches for repeatable tracking control tasks. Automatica. **37(7)**, 997–1006 (2001)
150. Xu, J.-X., Tan, Y.: A suboptimal learning control scheme for nonlinear systems with time-varying parametric uncertainties. J. Optimal Control – Applications and Theory. **22**, 111–126 (2001)
151. Xu, J.-X., Tan, Y.: Robust optimal design and convergence properties analysis of iterative learning control approaches. Automatica. **38(11)**, 1867–1880 (2002)
152. Xu, J.-X., Tan, Y.: A composite energy function based learning control approach for nonlinear systems with time-varying parametric uncertainties. IEEE Trans. Automatic Control. **47**, 1940–1945 (2002)
153. Xu, J.-X., Tan, Y.: Linear and Nonlinear Iterative Learning Control. Lecture Notes in Control and Information Science, **291**. Springer-Verlag, Germany, ISBN 3-540-40173-3 (2003)
154. Xu, J.-X., Tan, Y., Lee, T.H.: Iterative learning control design based on composite energy function with input saturation. Automatica. **40(8)**, 1371–1377 (2004)
155. Xu, J.-X., Xu, J.: On iterative learning for different tracking tasks in the presence of time-varying uncertaintie. IEEE Trans. Systems, Man, and Cybernetics, Part B. **34(1)**, 589–597 (2004)
156. Xu, J.-X., Tan, Y., Lee, T.H.: Iterative learning control design based on composite energy function with input saturation. Automatica. **40(8)**, 1371–1377 (2004)
157. Xu, J.-X., Panda, S.K., Pan, Y.J., Lee, T.H., Lam, B.H.: A modular control scheme for PMSM speed control with pulsating torque minimization. IEEE Trans. Industrial Electronics, **51(3)**, 526–536 (2004)
158. Xu, J.-X., Yan, R.: On initial conditions in iterative learning control. IEEE Trans. Automatic Control. **50(9)**, 1349–1354 (2005)
159. Xu, J.-X., Xu, J., Lee, T.H.: Iterative learning control for systems with input deadzone. IEEE Trans. Automatic Control, **50(9)**, 1455–1459 (2005)
160. Xu, J.-X., Yan, R.: On repetitive learning control for periodic tracking tasks. IEEE Trans. Automatic Control. **51(11)**, 1842–1848 (2006)
161. Yang, D.R., Lee, K.S., Ahn, H.J., Lee, J.H.: Experimental application of a quadratic optimal iterative learning control method for control of wafer temperature uniformity in rapid thermal processing. IEEE Trans. Semiconductor Manufacturing. **16(1)**, 36C44 (2003)
162. Yoshizawa, T.: Stability Theory by Liapunov's Second Method. Mathematical Society of Japan. Tokyo (1975)
163. Zhuang, H.Q., Roth, Z.S., Robot calibration using the CPC error model. Robotics and Computer-Integrated Manufacturing. **9(3)**, 225–237 (1992)
164. Ziegler, J.B., Nichols, N.B.: Optimum settings for automatic controllers. Trans. Amer. Soc. Mech. Eng. **64**, 759–768 (1942)
165. Zlatanov, D., Fenton, R.G., Benhabib, B.: Identification and classification of the singular configurations of mechanisms. Mechanism and Machine Theory. **33(6)**, 743–460 (1998)

Index

Add-on controller, 132
Alignment condition, 20, 22
Alumina powder, 97
Anti-noise filter, 75
Auto-tuning, 141, 157

Backlash, 4, 54
Band pass filter, 15
Band-pass filters, 30
Batch operations, 66
Batch process/reactor, 80, 87, 95, 97
Benchmark, 154
Bode plot, 11
Butterworth filter, 25

Calibration, 5, 169
Cascade ILC, 12
Cascade structure, 4, 12, 124
CCL, 4, 10, 23, 26, 29, 31, 37, 107, 108
Chemical reactor, 65
Closed-chain, 169, 171
Co-energy, 122
Complementary sensitivity function, 87
Concentration control, 65
Coupled tank, 161
Current controller, 110, 128
Current limits, 34
Current offset, 104
Current scaling error, 105

DC motor, 30
Dead time, 67, 87
Deadzone, 4, 41, 50, 55
Distortion, 15
Drop-rise phenomenon, 23

Electrical drive, 3, 101, 121

Embedded structure, 12
Exhaustive searching, 153
Extremum seeking, 156

Feasibility condition, 53
Feasible solutions, 36, 44
Feed-forward initialization, 79
Filter design, 7
Filter-based ILC, 66
Filtering, 24
Fixed initial shift, 20, 21
Flux fringing, 122
Flux harmonics, 104
Flux linkage, 130
Flux-linkage, 103
FOPDT, 65–67, 161
foreword, ix
Forgetting factor, 20, 24, 25, 108, 110
Fourier series based learning, 108
Freeway density, 54
Furnace reactor, 85

Genetic algorithm, 36
Geometric rate, 21
GLC, 17, 51
Gradient information, 150

Heat exchange, 5
Heat exchanger, 68
Hysteresis controller, 124

i.i.c., 19, 51, 55, 58, 81
Identification, 5, 169
Ignorance function, 89
ILC servomechanism, 30
Incremental cascade ILC, 14
Incremental inductance, 131
Incremental structure, 15

Indirect torque controller, 123
Input constraints, 53
Input non-linearities, 47
Integrated square error, 145
Integrated time-weighted square error, 145
Intelligent control, 30
Internal model, 1
Internal model control, 157
Inverse kinematics, 171, 176
Iterative feedback tuning, 156
Iterative learning identification, 169
Iterative learning tuning, 141, 150
Iterative searching, 153

Jacobian, 17, 18

Learning convergence speed, 30, 34, 41
Least Squares, 33, 42, 131
Level segment, 79
Low gain feedback, 45
LPF, 15, 24, 25, 76
Lyapunov–Krasovskii functional, 22

Magnetic saturation, 122
Manipulator, 170
Max-min operation, 43
Mean value theorem, 18
Memory arrays, 8, 11
Memoryless, 49
microrobot, 169
Model predictive ILC, 4
Modeling accuracy, 30
Motion control, 3
Motor torque, 104
Moving average, 73
Multiplicative perturbations, 85

Newton method, 175
Nominal model, 32, 42
Non-affine-in-input, 42, 47
Non-integer harmonics, 116
Non-minimum phase, 27
Non-repeatable, 23, 30
Non-smooth non-linearities, 63
Non-symmetric deadzone, 58
Nyquist curve, 75, 77
Nyquist frequency, 30, 34

Objective function, 4, 5, 29, 34, 36, 38, 39, 43,
    143, 145, 149, 155
Optimal CCL, 37
Optimal PCCL, 40
Optimal PCL, 35
Optimal tuning, 5

Optimization, 144
Overshoot, 77, 145

PCCL, 4, 11, 26, 29, 31, 40, 59
PCL, 4, 8, 14, 24, 26, 29, 31, 35, 107
Periodic learning, 22
Periodic oscillations, 102
Permanent magnetic synchronous motor
    (PMSM), 5, 101
Pharmaceutic industry, 65
Phase torque, 122
PID, 4, 5, 22, 34, 35, 38, 58, 65, 67, 68, 80, 87,
    98, 110, 141, 157
PID type ILC, 4
Piezo motor, 3, 55
Plant/model uncertainty, 85, 88
Polynomial, 33, 131
Process control, 3, 4
Process gradient, 148
Progressive initial condition, 20
PWM, 32

Ramp segment, 79
Random initial condition, 20
Re-programming, 13
Recursion equation, 89, 94
Relative degree, 24, 27, 35
Relay tuning, 98
Repetitive control, 22, 23, 107
Repetitive control tasks, 7
Repetitive learning control, 22
Rise time, 145
Robot, 171
Robust ILC, 3
Robust optimal design, 43
Robust optimal ILC, 4
Robust optimal PCCL, 42
Robustness, 7, 20, 23, 29, 34, 37, 175

Sampled-data, 25, 31, 129, 160
Sampling delay, 26
Sampling effect, 7
Saturation, 4, 53
Self-adaptation, 151
Sensitivity function, 87
Servo, 3, 29, 47
Setpoint tracking, 63
Settling time, 22, 145
Smith predictor, 5, 85, 88
Speed controller, 110
Speed regulation, 101
Speed ripple, 101, 106, 110
Speed ripple factor, 112
SRM, 5
Steepest descending, 151

Stopping criterion, 158
Switched reluctance motor, 121
System repeatability, 58

Temperature control, 65, 75, 95, 99
Thermocouple, 97
Thyristor, 97
Time delay, 86, 94
Torque estimation, 130
Torque harmonics, 104
Torque pulsation, 101, 104
Torque ripple, 121, 123, 133
Torque sharing function (TSF), 125

Torque transducer, 102, 110

Variable speed drive, 122

Wafer industry, 65
water-heating process, 76, 82
Worst case optimal performance, 45

X-Y table, 4, 21, 29

Zero-phase filter, 72, 75
Ziegler–Nichols tuning, 155
Ziegler-Nichols tuning, 68

Other titles published in this series (continued):

Soft Sensors for Monitoring and Control of Industrial Processes
Luigi Fortuna, Salvatore Graziani,
Alessandro Rizzo and Maria G. Xibilia

Adaptive Voltage Control in Power Systems
Giuseppe Fusco and Mario Russo

Advanced Control of Industrial Processes
Piotr Tatjewski

Process Control Performance Assessment
Andrzej W. Ordys, Damien Uduehi and
Michael A. Johnson (Eds.)

Modelling and Analysis of Hybrid Supervisory Systems
Emilia Villani, Paulo E. Miyagi and
Robert Valette

Process Control
Jie Bao and Peter L. Lee

Distributed Embedded Control Systems
Matjaž Colnarič, Domen Verber and
Wolfgang A. Halang

Precision Motion Control (2nd Ed.)
Tan Kok Kiong, Lee Tong Heng and
Huang Sunan

Optimal Control of Wind Energy Systems
Iulian Munteanu, Antoneta Iuliana Bratcu,
Nicolaos-Antonio Cutululis and Emil
Ceangă

Identification of Continuous-time Models from Sampled Data
Hugues Garnier and Liuping Wang (Eds.)

Model-based Process Supervision
Arun K. Samantaray and
Belkacem Bouamama

Diagnosis of Process Nonlinearities and Valve Stiction
M.A.A. Shoukat Choudhury, Sirish L.
Shah, and Nina F. Thornhill

Magnetic Control of Tokamak Plasmas
Marco Ariola and Alfredo Pironti

Deadlock Resolution in Automated Manufacturing Systems
ZhiWu Li and MengChu Zhou
Publication due February 2009

Model Predictive Control Design and Implementation Using MATLAB®
Liuping Wang
Publication due March 2009

Drives and Control for Industrial Automation
Tan Kok Kiong and Andi S. Putra
Publication due May 2009

Design of Fault-tolerant Control Systems
Hassan Noura, Didier Theilliol,
Jean-Christophe Ponsart and Abbas
Chamseddine
Publication due June 2009

Control of Ships and Underwater Vehicles
Khac Duc Do and Jie Pan
Publication due June 2009

Dry Clutch Control for Automated Manual Transmission Vehicles
Pietro J. Dolcini, Carlos Canudas-de-Wit
and Hubert Béchart
Publication due June 2009

Stochastic Distribution Control System Design
Lei Guo and Hong Wang
Publication due June 2009

Predictive Functional Control
Jacques Richalet and Donal O'Donovan
Publication due June 2009

Advanced Control and Supervision of Mineral Processing Plants
Daniel Sbárbaro and René del Villar (Eds.)
Publication due July 2009

Fault-tolerant Flight Control and Guidance Systems
Guillaume Ducard
Publication due July 2009

Detection and Diagnosis of Stiction in Control Loops
Mohieddine Jelali and Biao Huang (Eds.)
Publication due October 2009

Active Braking Control Design for Road Vehicles
Sergio M. Savaresi and Mara Tanelli
Publication due November 2009